STEM CELLS: NUCLEAR REPROGRAMMING AND THERAPEUTIC APPLICATIONS

The Novartis Foundation is an international scientific and educational charity (UK Registered Charity No. 313574). Known until September 1997 as the Ciba Foundation, it was established in 1947 by the CIBA company of Basle, which merged with Sandoz in 1996, to form Novartis. The Foundation operates independently in London under English trust law. It was formally opened on 22 June 1949.

The Foundation promotes the study and general knowledge of science and in particular encourages international co-operation in scientific research. To this end, it organizes internationally acclaimed meetings (typically eight symposia and allied open meetings and 15–20 discussion meetings each year) and publishes eight books per year featuring the presented papers and discussions from the symposia. Although primarily an operational rather than a grant-making foundation, it awards bursaries to young scientists to attend the symposia and afterwards work with one of the other participants.

The Foundation's headquarters at 41 Portland Place, London W1B 1BN, provide library facilities, open to graduates in science and allied disciplines. Media relations are fostered by regular press conferences and by articles prepared by the Foundation's Science Writer in Residence. The Foundation offers accommodation and meeting facilities to visiting scientists and their societies.

Information on all Foundation activities can be found at http://www.novartisfound.org.uk

Novartis Foundation Symposium 265

STEM CELLS: NUCLEAR REPROGRAMMING AND THERAPEUTIC APPLICATIONS

2005

John Wiley & Sons, Ltd

Published in 2005 by John Wiley & Sons Ltd,
The Atrium, Southern Gate,
Chichester PO19 8SQ, UK

National 01243 779777
International (+44) 1243 779777
e-mail (for orders and customer service enquiries): cs-books@wiley.co.uk
Visit our Home Page on http://www.wileyeurope.com
or http://www.wiley.com

Other Wiley Editorial Offices

John Wiley & Sons Inc., 111 River Street, Hoboken, NJ 07030, USA

Jossey-Bass, 989 Market Street, San Francisco, CA 94103-1741, USA

Wiley-VCH Verlag GmbH, Boschstr. 12, D-69469 Weinheim, Germany

John Wiley & Sons Australia Ltd, 33 Park Road, Milton, Queensland 4064, Australia

John Wiley & Sons (Asia) Pte Ltd, 2 Clementi Loop #02-01, Jin Xing Distripark, Singapore 129809

John Wiley & Sons Canada Ltd, 22 Worcester Road, Etobicoke, Ontario, Canada M9W 1L1

Wiley also publishes its books in a variety of electronic formats. Some content that appears in print may not be available in electronic books.

Novartis Foundation Symposium 265
viii+221 pages, 11 figures, 5 tables

British Library Cataloguing in Publication Data

A catalogue record for this book is available from the British Library

ISBN-13 978-0-470-09143-2 (HB)

ISBN-10 0-470-09143-6 (HB)

Typeset in 10½ on 12½ pt Garamond by Dobbie Typesetting Limited, Tavistock, Devon.
Printed and bound in Great Britain by T. J. International Ltd, Padstow, Cornwall.
This book is printed on acid-free paper responsibly manufactured from sustainable forestry, in which at least two trees are planted for each one used for paper production.

Contents

Symposium on Stem cells: nuclear reprogramming and therapeutic applications, held at the Novartis Foundation, London, 2–4 March 2004.

Editors: Gregory Bock (Organizer) and Jamie Goode

This symposium is based on proposals made by Karin Blumer and Uwe Junker, and Adam Wilkins

Participants

Anders Björklund Division of Neurobiology, Lund University, Wallenberg Neuroscience Center, BMC A11, S-22184 Lund, Sweden

Karin Blumer Novartis Corporate Policy, WSJ 200.330, CH-4002 Basel, Switzerland

Martin Evans Cardiff School of Biosciences, Biomedical Sciences Building, Museum Avenue, PO Box 911, Cardiff CF10 3US, UK

John Gearhart (*Chair*) Department of Gynaecology and Obstetrics, The Johns Hopkins School of Medicine, Baltimore, MD 21287, USA

Lior Gepstein Cardiovascular Research Laboratory, The Bruce Rappaport Faculty of Medicine, POB 9649, Haifa 31096, Israel

Steven A. Goldman Department of Neurology, University of Rochester Medical Center, 601 Elmwood Ave./MRBX, Box 645, Rochester, NY 14642, USA

Markus Grompe Department of Molecular and Medical Genetics, Oregon Health & Science University, 3181 S.W. Sam Jackson Park Rd, Portland, OR 97239-3098, USA

John B. Gurdon Wellcome Trust/Cancer Research UK Gurdon Institute, Tennis Court Road, Cambridge CB2 IQN, UK

Rudolf Jaenisch Whitehead Institute, Nine Cambridge Center, Cambridge, MA 02142-1479, USA

Uwe Junker Novartis Institutes for Biomedical Research, Arthritis and Bone Metabolism Biology 1, One Health Plaza 436/4014, East Hanover, NJ 07936, USA

Robin Lovell-Badge Division of Developmental Genetics, MRC National Institute for Medical Research, The Ridgeway, Mill Hill, London NW7 1AA, UK

Stephen Minger Stem Cell Biology Laboratory, Centre for Neuroscience Research, King's College London, Guy's Campus/Hodgkin Building, London SE1 1UL, UK

Thomas Murray The Hastings Center, 21 Malcolm Gordon Road, Garrison, NY 10524-5555, USA

Bernardo Nadal-Ginard Cardiovascular Research Institute, Department of Medicine, New York Medical College, Valhalla, NY 10595, USA

Davor Solter Department of Developmental Biology, Max Planck Institute of Immunobiology, Stubeweg 51, D-79011 Freiburg, Germany

Bernat Soria Institute of Bioengineering and Department of Physiology, School of Medicine, Miguel Hernandez University, Alicante, Spain

Stephen Sullivan (*Novartis Foundation Bursar*) The Surani Lab, Wellcome Trust/Cancer Research UK Institute, Tennis Court Road, University of Cambridge, Cambridge CB2 1QR, UK

Alan Trounson Monash Immunology and Stem Cell Labs, Department of Physiology, Monash University, Wellington Road, Clayton, Victoria 3800, Australia

Catherine Verfaillie Department of Medicine, University of Minnesota, Moos Tower 14-287A, Mayo Mail Code 716, 420 Delaware St. SE, Minneapolis, MN 55455, USA

Irving Weissman Department of Pathology, Stanford University Medical Center, B257 Beckman Center, Stanford, CA 94305, USA

Adam Wilkins BioEssays, 10/11 Tredgold Lane, Napier Street, Cambridge CB1 1HN, UK

Ian Wilmut Roslin Institute, Roslin BioCentre, Midlothian, EH25 9PS, UK

Chair's introduction

John Gearhart

Department of Gynaecology and Obstetrics, The Johns Hopkins School of Medicine, Baltimore, MD 21287, USA

In my introduction I would like to frame some of the questions that we will be addressing over the next several days. Some of these are obvious, but I hope that out of this meeting we will synthesize some new ones.

Clearly, we would like to know what stem cells are, and how we can define them functionally and molecularly. What are their properties? What is this entity known as 'stemness' that has now appeared in the literature? Are we talking about epigenetics or chromatin structure? All of these are likely to be involved. What are the sources of stem cells? In fetal tissues, embryonic tissues, adult tissues, how do we recognize, isolate, characterize and grow them? These are all issues that will be central to our discussions over the next three days. Most importantly, we are concerned about how we can control these cells in the laboratory. How do we get them to do the things that we want them to do, such as differentiating into specific cell types with high efficiency? What strategies are currently used and how successful are these? In grafts, can we get stem cells or their derivatives to do what we want them to do, for example differentiating in a tissue appropriate manner, with no migration and no tumour formation? Also, it is clear that stem cells exist *in situ*. Will it be possible to manipulate them *in situ* without taking them out and then grafting them back? Somatic cell nuclear transfer (SCNT, or cell nuclear transfer, CNT) is an important technology in stem cell research. Embryonic stem cells derived from patients will be important in avoiding the immune response to grafted cells and in studying diseases. SCNT will also enable us to study and therefore gain an understanding of the basis of cell differentiation.

Then there is the whole issue of the development of cell-based therapies. While there are some published proofs of principle, there remain many obstacles to developing safe and effective cell-based therapies. Some believe that there should be a more complete understanding of the pathogenesis of the disease and of injuries, so we know exactly what we are doing when we go in and try to replace diseased, injured or dead cells. One area that should receive more attention in the public discussions is the issue of time-frames for the development of safe and effective therapies. We are faced with an issue in the USA of politicians who are elected for 2, 4 or 6 years and want something done within that time-frame. We must be circumspect in making predictions about time-frames. The expectations that go with these predictions or promises that are not met will have serious repercussions. At this meeting we have Tom Murray, who refers to himself as our token ethicist. There are many ethical issues in this field that need to be dealt with fairly. All of us as scientists would feel more comfortable if suitable guidelines were in place. This will come from discussions which not only reflect the sensitivity of the issues, but would also permit robust forward movement of our research. In the USA we see the increasing politicization of the stem cell debate: two members of the President's Council on Bioethics were recently replaced because their views were not in sync with those of the chairman of the Council. We do not know how this will affect recommendations for national policy, but it results in less credibility for the Council.

I am sure that everyone in this room will contribute something significant over the next few days, and we look forward to this mixing of scientists which is at the heart of these symposia.

Many of us have heroes in embryology. One person who stands out for me is Karl Ernst von Baer, and I wanted to reflect on a paraphrase of what he wrote more than 100 years ago: 'All new and truly important ideas and discoveries must pass through three stages: first, dismissed as nonsense, then rejected as against religion, and finally acknowledged as true, with the proviso from initial opponents that they knew it all along.' We are currently between stages two and three.

What is a stem cell?

Davor Solter

Max-Planck Institute of Immunobiology, Stübeweg 51, 79108 Freiburg, Germany

Abstract. The nature, origin, sources and possible modes of derivation of human embryonic stem cells are scrutinized. Can cells, with a capacity for differentiation equal to that of embryonic stem cells, be derived from adult organisms and what are the appropriate sources for such cells? Do stem cells found in adult organs and tissues possess developmental plasticity, i.e. are they able to transdifferentiate across germ layer boundaries? What are the anticipated short- and long-term uses of embryonic and adult stem cells? The current state of science and contentious issues as related to these questions are discussed.

2005 Stem cells: nuclear reprogramming and therapeutic applications. Wiley, Chichester (Novartis Foundation Symposium 265) p 3–19

Stem cells can proliferate (maybe indefinitely) without changing their phenotype, but they are also able to simultaneously change (differentiate) into one or more new cell types. This transition to the final differentiated phenotype can either be direct or proceed in several well-defined steps. It is not entirely clear whether it is essential that division must precede the appearance of the differentiated phenotype or whether a stem cell can change its phenotype without dividing. Neither is it clear whether this division always results in two differentiated daughter cells or one daughter and one stem cell. It is very likely that both during development and in the adult organism these two types of division take place.

The literature on stem cells is vast, but in the following I will concentrate predominantly on the subject of embryonic stem (ES) cells (with the emphasis on human) and also on some recent insights into the use of adult stem cells in cell and tissue replacement therapies. This subject has attracted tremendous attention in recent years, the number of publications is overwhelming and I apologize to all authors whose work I could not cite due to lack of space. For this reason I will limit my presentation to the discussion of only a few examples which will hopefully illustrate the underlying principles. A more detailed conceptual analysis and numerous helpful references can be found in some of the recent reviews and books on the subject (Blau et al 2001, Fuchs & Segre 2000, Solter et al 2003, Weissman 2000). Finally, it has to be mentioned that the field is mired in controversy to a degree well beyond what is usual in scientific endeavours. The rush to publish, probably due to an altruistic desire to help incurably afflicted patients but also spurred by pecuniary motives, has resulted in the repeated publication of entirely opposite and irreconcilable results. For this reason several important issues dealing with basic stem cell biology still remain completely open.

ES cells

The concept of ES cells arose slowly out of pioneering work with mouse and human teratocarcinomas (Evans & Hunter 2005, this volume, Evans & Kaufman 1981). Several decades ago it became apparent that these rare tumours contain cells, called embryonal carcinoma cells, the stem cells of the tumour which are also able to differentiate into a multitude of adult cells and tissues found in these malignancies. The observation that one can easily obtain teratocarcinomas from embryos transferred to extrauterine sites suggested that it should be possible to obtain ES cells with similar properties directly from the embryos. This approach may have been obvious, but it took some time before the first ES cells were derived from mouse blastocysts (Evans & Kaufman 1981, Martin 1981). The derivation of human ES cells (Thomson et al 1998) came several years later, most likely due to difficulties in securing a sufficient number of healthy embryos for experimentation. There may be further biological and technical obstacles which make the isolation of ES cells difficult or impossible since, despite many efforts, *bona fide* ES cells have not yet been isolated from the embryos of farm animals (cow, sheep) or rats.

The derivation of mouse ES cells (also human) requires medium containing serum and the presence of fibroblastic feeder layers (Evans & Kaufman 1981, Martin 1981, Thomson et al 1998). Only recently did it become possible to isolate and maintain mouse ES cells without serum and feeder in a medium containing leukaemia inhibitory factor (LIF) and bone morphogenetic protein 4 (BMP4) (Ying et al 2003). It would clearly be very advantageous if human ES cells intended for clinical use were derived in a similar manner in order to avoid any danger of contamination with endogenous retroviruses. The growth (but not derivation) of human ES cells using human fibroblast feeder has been reported (Richards et al 2002) as well as attempts to identify factors which would keep them in an undifferentiated state (Sato et al 2004). Unfortunately, even the most recent derivation of numerous human ES cell lines involved mouse embryo fibroblasts as a feeder layer and human serum protein (Cowan et al 2004).

Once derived, ES cells have to be defined and characterized, which has been accomplished in great detail for mouse ES cells. The most important and stringent test of mouse ES cells is their ability to colonize a normal blastocyst after injection, resulting in chimeric organisms whose many tissues and, most importantly, germ cells are derived from injected ES cells. Human ES cells cannot be tested this way, thus attempts have been made to define the appropriate criteria which could and should be applied to freshly isolated cell lines (Brivanlou et al 2003). These include morphological criteria, the ability to differentiate *in vivo* and *in vitro*, behaviour in culture and the presence or absence of specific diagnostic markers. Essentially all markers currently in use were not developed for the purpose of identifying human ES cells but were discovered during studies of mouse early development, and of mouse and human teratocarcinomas (Brivanlou et al 2003, Cowan et al 2004, Henderson et al 2002). Recently two groups (Ivanova et al 2002, Ramalho-Santos et al 2002) tried a more systematic approach using gene expression profiling of several mouse stem cells (embryonic, haematopoietic and neural) in order to define the genes commonly up-regulated in stem cells. Each group presented a list of 200–300 'stemness' genes and suggested that they define the molecular signature of stem cells. Unfortunately the overlap between the two reported sets was negligible (Evsikov & Solter 2003, Fortunel et al 2003), thus the technology is either not yet suitable for this type of analysis or a distinct set of genes characteristic for all stem cells does not exist. Moreover the overlap of sets of genes up-regulated in mouse ES cells as presented by three different groups was around 20% (Fortunel et al 2003, Ivanova et al 2002, Ramalho-Santos et al 2002), which suggests that defining ES cells themselves by their gene expression profile may be difficult or unrealistic. Recent similar work using human ES cells (Sato et al 2003, Sperger et al 2003) further emphasizes the problem. A comparison of the lists of the most significantly up-regulated genes from both groups shows only one gene common to both, namely *POU5F1* (better known as Oct3/4), the gene whose unique expression in ES cells has been known for years. The clinical application of ES cell derivatives will require reliable markers for the characterization and, even more importantly, for the separation of different cell populations, and the search for such markers should be one of the highest priorities.

The usefulness of human ES cells in a clinical setting is predicated upon their ability to differentiate *in vitro*. It has long been known that mouse ES cells, and for that matter, mouse and human embryonal carcinoma (EC) cells possess a remarkable capacity to differentiate *in vitro*. However, in most cases differentiation was spontaneous and not directed into one specific cell type. Differentiation usually required the formation of semi-organized clumps of ES cells known as embryoid bodies (see Evans & Hunter 2005, this volume), which, upon attachment, grew into a disorganized mixture of many adult cell types. This type of differentiation, though interesting in its basic biology, is not very useful for clinical application and methods will have to be developed to direct the differentiation of ES cells along predetermined and desired pathways. Recent reports demonstrate that one can use currently available knowledge about tissue-determining factors and cellular interactions during normal development to design protocols which allow the differentiation of ES cells *in vitro* (Loebel et al 2003).

Such examples include the derivation from mouse ES cells of functional motor neurons (Wichterle et al 2002), dopaminergic neurons (Kim et al 2002), multilayered keratinocytes forming embryonic skin (Coraux et al 2003) and functional hair cells capable of colonizing cochlear or vestibular epithelium of the inner ear (Li et al 2003). The described protocols are relatively complex and none of them is fully successful, i.e. not all ES cells differentiate into the desired population and it is always necessary to sort out specific cells, but these results certainly indicate that ES cells can give rise to clinically applicable products. Human ES cells are equally capable of differentiation *in vitro* (Carpenter et al 2003) and experiments designed to determine which factors can be used to direct their differentiation are under way (Schuldiner et al 2000).

For the clinical application of ES cell derivatives it is essential that the initial population is karyotypically normal and stable. While some reports suggested that human ES cells can be cultured for an extensive period of time and remain diploid (Buzzard et al 2004, Thomson et al 1998), others found evidence of aneuploidy, mostly trisomy of chromosome 12 and sometimes trisomy 17 (Draper et al 2004). It is possible that these chromosomal abnormalities impart growth advantage and aneuploid cells eventually overgrow the cells with a normal chromosomal complement. It is unclear how common the phenomenon is and how much it will affect the clinical use of ES cells, but it certainly warrants attention.

The making of ES cells

ES cells, mouse and human, have traditionally been derived from inner cell mass cells of the blastocyst (Evans & Kaufman 1981, Martin 1981, Thomson et al 1998). While these cells are perfectly adequate for basic biological research and for developing technologies for clinical application, it would be beneficial for many reasons if ES cells, whose derivatives are intended for use in cell and tissue replacement therapies, were genetically identical to the host, i.e. autologous (Solter 1999, Solter & Gearhart 1999, Solter et al 2003). To achieve this goal it would be necessary to turn the patient's adult cell into an ES cell, and the conceptually simplest solution would be the procedure known as 'therapeutic cloning'. Briefly this involves transferring an adult cell nucleus of a person in need of ES cells into an enucleated oocyte, allowing this oocyte to develop to blastocyst, and deriving an ES cell line from this blastocyst using standard protocols. All the necessary technology is available and such ES cell lines have been derived from adult mouse cells (Wakayama et al 2001). Very recently the same approach was used to derive a human ES cell line (Hwang et al 2004) and, though some questions remain, there is little doubt that such simple therapeutic cloning is possible. Nevertheless, it is unlikely that therapeutic cloning as described above will become a long-term solution once the utility of ES cells in clinical practice is indeed established.

Broadly speaking, two sets of problems, one technical and the other social (moral, religious, philosophical) present obstacles (Solter et al 2003). The procedure is technically very demanding, time-consuming and extremely inefficient. It also requires the destruction of a human embryo possessing a very limited but real possibility of further development, which makes the technique illegal in large parts of the world (Solter et al 2003). What are the alternatives? One could use non-human, mammalian oocytes which would eliminate the demand for human oocytes, though these may be more readily available in the future if the derivation of functional oocytes *in vitro* from ES cells proves possible, as a recent report has indicated (Hübner et al 2003). One example of the derivation of ES cell lines from blastocysts following the transfer of human somatic cell nuclei into a rabbit oocyte has been described (Chen et al 2003), however this probably does not resolve the above-mentioned objections. The ideal solution would obviously be to bypass the entire process of nuclear transfer and to develop protocols, which would enable us to reprogramme an adult somatic cell into ES cells in culture using a battery of growth and differentiation factors. Although still in very preliminary stages, this approach is likely to succeed eventually and become one of the standard ways of producing ES cells tailored for a specific individual (Dennis 2003, Park et al 2003).

Are there functional alternatives to ES cells, i.e. can one make cells not derived from a blastocyst but which possess an equal capacity for differentiation? The best known alternatives are so-called embryonic germ (EG) cells derived from primordial germ cells localized in the genital ridges. Both mouse and human EG cells have been isolated and characterized (Matsui et al 1992, Shamblott et al 1998) and they have properties very similar to those of ES cells. Although they do not invoke such intense moral opprobrium as ES cells, the technical problems involving their use are similar, if not worse. Using the capacity to make chimeras following injection into the blastocyst as a criterion, several other cell types have been described. Such stem cells have been derived from the brain (Clarke et al 2000), epidermis (Liang & Bickenbach 2002) and bone marrow mesenchyme (Jiang et al 2002). Although interesting, none of these results have been independently confirmed and in no case was germ line chimerism observed. Thus, although it is theoretically possible that cells with a capacity similar to that of ES cells can be isolated from adults, this has not been unequivocally demonstrated.

Adult stem cells — developmental plasticity

Every multicellular organism loses an innumerable number of cells every day and, in order to continue functioning, these cells have to be constantly replaced. The presence of stem cells unique to a given organ was demonstrated a long time ago and their function and biology have since been extensively studied and described (Blau et al 2001, Fuchs & Segre 2000, Solter et al 2003, Weissman 2000).

It is well beyond the scope of this article to review this vast and complex field and I will only briefly discuss one recent issue, namely the developmental plasticity of adult stem cells. The general assumption was that adult stem cells have a limited capacity to differentiate. They can produce many cell types, e.g. haematopoietic stem cells giving rise to all blood elements (erythrocytes, leukocytes, lymphocytes, macrophages) or neural stem cells giving rise to neurons and glia, but they were not expected to cross the germ layer barrier, i.e. neuronal stem cells were not supposed to produce mesodermal derivatives like blood and vice versa. This notion of limited potential was recently challenged by numerous reports (reviewed in Solter et al 2003, Wagers & Weissman 2004). These include the description of bone marrow stem cells giving rise to neuronal cells (Brazelton et al 2000, Mezey et al 2000), lung and skin epithelium (Krause et al 2001), liver (Körbling et al 2002, Lagasse et al 2000) and myocardium (Orlic et al 2001). The reverse direction of transdifferentiation was also described, i.e. muscle stem cells (Jackson et al 1999) or neural stem cells (Bjornson et al 1999) giving rise to haematopoietic stem cells. These results were subsequently contested on several grounds (Wagers & Weissman 2004) including claims that detailed analysis did not find any evidence that haematopoietic stem cells transdifferentiated into neural cells (Castro et al 2002, Wagers et al 2002), liver (Kanazawa & Verma 2003) or myocardium (Balsam et al 2004, Murry et al 2004). It has been suggested that most, if not all, examples of transdifferentiation can be explained by a fusion of donor and host cells (Alvarez-Dolado et al 2003, Weimann et al 2003). This fusion product would then display markers of host cells and differentiated markers of donor cells, which, in the absence of further analysis, could deceive investigators into assuming that transdifferentiation took place. Is fusion the entire explanation and should all claims of transdifferentiation be discounted as interpretation errors? It is probably too early to tell. Cogle et al (2004) analysed brain tissue in autopsy material from patients who received bone-marrow transplantation as part of anti-cancer treatment and observed donor cells (without any evidence for cell fusion) among neurons and glia, especially if death occurred several years after transplantation. Similarly, Ianus et al (2003) reported the presence of functional pancreatic β cells following bone marrow transplantation, and again cell fusion was not observed. In these cases unfractionated bone marrow cells were used, and it is thus possible that circulating stem cells from other organs were injected. Obviously a strict characterization of injected cells would be one of several necessary criteria before transdifferentiation can be accepted as a true biological phenomenon (Goolsby et al 2003, Wagers & Weissman 2004). Adult stem cells, provided they can be taken from the patient, would avoid problems associated with immune rejection and their isolation, and their use does not present any ethical problem of the kind associated with ES cells. However, beside the unresolved issue of transdifferentiation (which, if it exists, may be by itself a mixed blessing) several other problems (ease of recovery, potential for expansion, somatic mutation, etc.) will have to be resolved before their extensive medical application can be considered.

Use of ES cells

Current debates about the potential use of ES cells and their differentiated derivatives centre on their clinical use in cell, tissue and possibly organ replacement. This may not be the most important and beneficial application in the long run. Stem cells can be transfected with a specific DNA and thus serve as a vehicle for gene therapy, thus avoiding the unreliability and danger of using viral vectors. Moreover, homologous recombination can be used to correct the genetic defect in the patient's own ES cells and their differentiated derivatives employed to eliminate the mutant phenotype (Rideout et al 2002, Solter 1999). The isolation of ES cells from patients with genetic diseases (by nuclear transfer or other means discussed above) will also provide us with *in vitro* models of these diseases and enable us to study their pathogenesis. In addition, such cells would serve as valuable models for drug discovery and testing. Most of the uses of ES cells lie in the future and, though very attractive, the benefits are by no means ensured. Poorly informed ethical squabbles, injudicious and premature claims of great health benefits and the publication of poorly controlled, rushed experiments threaten to inundate and seriously damage this otherwise very promising area of research and we should make every effort to protect it.

References

Alvarez-Dolado M, Pardal R, Garcia-Verdugo JM et al 2003 Fusion of bone-marrow-derived cells with Purkinje neurons, cardiomyocytes and hepatocytes. Nature 425:968–973

Balsam LB, Wagers AJ, Christensen JL, Kofidis T, Weissman IL, Robbins RC 2004 Haematopoietic stem cells adopt mature haematopoietic fates in ischaemic myocardium. Nature 428:668–673

Bjornson CRR, Rietze RL, Reynolds BA, Magli MC, Vescovi AL 1999 Turning brain into blood: a hematopoietic fate adopted by adult neural stem cells in vivo. Science 283:534–537

Blau HM, Brazelton TR, Weimann JM 2001 The evolving concept of a stem cell: entity or function? Cell 105:829–841

Brazelton TR, Rossi FMV, Keshet GI, Blau HM 2000 From marrow to brain: expression of neuronal phenotypes in adult mice. Science 290:1775–1779

Brivanlou AH, Gage FH, Jaenisch R, Jessell T, Melton D, Rossant J 2003 Setting standards for human embryonic stem cells. Science 300:913–916

Buzzard JJ, Gough NM, Crook JM, Colman A 2004 Karyotype of human ES cells during extended culture. Nat Biotechnol 22:381–382

Carpenter MK, Rosler E, Rao MS 2003 Characterization and differentiation of human embryonic stem cells. Cloning Stem Cells 5:79–88

Castro RF, Jackson KA, Goodell MA, Robertson CS, Liu H, Shine HD 2002 Failure of bone marrow cells to transdifferentiate into neural cells in vivo. Science 297:1299

Chen Y, He ZX, Liu A et al 2003 Embryonic stem cells generated by nuclear transfer of human somatic nuclei into rabbit oocytes. Cell Res 13:251–263

Clarke DL, Johansson CB, Wilbertz J et al 2000 Generalized potential of adult neural stem cells. Science 288:1660–1663

Cogle CR, Yachnis AT, Laywell ED et al 2004 Bone marrow transdifferentiation in brain after transplantation: a retrospective study. Lancet 363:1432–1437

Coraux C, Hilmi C, Rouleau M et al 2003 Reconstituted skin from murine embryonic stem cells. Curr Biol 13:849–853

Cowan CA, Klimanskaya I, McMahon J et al 2004 Derivation of embryonic stem-cell lines from human blastocysts. N Engl J Med 350:1353–1356

Dennis C 2003 Take a cell, any cell. Nature 426:490–491

Draper JS, Smith K, Gokhale P et al 2004 Recurrent gain of chromosomes 17q and 12 in cultured human embryonic stem cells. Nat Biotechnol 22:53–54

Evans MJ, Kaufman MH 1981 Establishment in culture of pluripotential cells from mouse embryos. Nature 292:154–156

Evans MJ, Hunter S 2005 Embryonic stem cells: a perspective. In: Stem cells: nuclear reprogramming and therapeutic applications (Novartis Found Symp 265), p 98–106

Evsikov AV, Solter D 2003 Comment on 'stemness': transcriptional profiling of embryonic and adult stem cells and a stem cell molecular signature. Science 302:393

Fortunel NO, Otu HH, Ng HH et al 2003 Comment on'stemness': transcriptional profiling of embryonic and adult stem cells and a stem cell molecular signature. Science 302:393

Fuchs E, Segre JA 2000 Stem cells: a new lease on life. Cell 100:143–155

Goolsby J, Marty MC, Heletz D et al 2003 Hematopoietic progenitors express neural genes. Proc Natl Acad Sci USA 100:14926–14931

Henderson JK, Draper JS, Baillie HS et al 2002 Preimplantation human embryos and embryonic stem cells show comparable expression of stage-specific embryonic antigens. Stem Cells 20:329–337

Hübner K, Fuhrmann G, Christenson LK et al 2003 Derivation of oocytes from mouse embryonic stem cells. Science 300:1251–1256

Hwang WS, Ryu YJ, Park JH et al Y 2004 Evidence of a pluripotent human embryonic stem cell line derived from a cloned blastocyst. Science 303:1669–1674

Ianus A, Holz GG, Theise ND, Hussain MA 2003 In vivo derivation of glucose-competent pancreatic endocrine cells from bone marrow without evidence of cell fusion. J Clin Invest 111:843–850

Ivanova NB, Dimos JT, Schaniel C, Hackney JA, Moore KA, Lemischka IR 2002 A stem cell molecular signature. Science 298:601–604

Jackson KA, Mi T, Goodell MA 1999 Hematopoietic potential of stem cells isolated from murine skeletal muscle. Proc Natl Acad Sci USA 96:14482–14486

Jiang Y, Jahagirdar BN, Reinhardt RL et al 2002 Pluripotency of mesenchymal stem cells derived from adult marrow. Nature 418:41–49

Kanazawa Y, Verma IM 2003 Little evidence of bone marrow-derived hepatocytes in the replacement of injured liver. Proc Natl Acad Sci USA 100:11850–11853

Kim J-H, Auerbach JM, Rodríguez-Gómez JA et al 2002 Dopamine neurons derived from embryonic stem cells function in an animal model of Parkinson's disease. Nature 418:50–56

Körbling M, Katz RL, Khanna A et al 2002 Hepatocytes and epithelial cells of donor origin in recipients of peripheral-blood stem cells. N Engl J Med 346:738–746
Krause DS, Theise ND, Collector MI et al 2001 Multi-organ, multi-lineage engraftment by a single bone marrow-derived stem cell. Cell 105:369–377
Lagasse E, Connors H, Al-Dhalimy M et al 2000 Purified hematopoietic stem cells can differentiate into hepatocytes *in vivo*. Nat Med 6:1229–1234
Li H, Roblin G, Liu H, Heller S 2003 Generation of hair cells by stepwise differentiation of embryonic stem cells. Proc Natl Acad Sci USA 100:13495–13500
Liang L, Bickenbach JR 2002 Somatic epidermal stem cells can produce multiple cell lineages during development. Stem Cells 20:21–31
Loebel DAF, Watson CM, De Young RA, Tam PPL 2003 Lineage choice and differentiation in mouse embryos and embryonic stem cells. Dev Biol 264:1–14
Martin GR 1981 Isolation of a pluripotent cell line from early mouse embryos cultured in medium conditioned by teratocarcinoma stem cells. Proc Natl Acad Sci USA 78:7634–7638
Matsui Y, Zsebo K, Hogan BLM 1992 Derivation of pluripotential embryonic stem cells from murine primordial germ cells in culture. Cell 70:841–847
Mezey É, Chandross KJ, Harta G, Maki RA, McKercher SR 2000 Turning blood into brain: cells bearing neuronal antigens generated in vivo from bone marrow. Science 290:1779–1782
Murry CE, Soonpaa MH, Reinecke H et al 2004 Haematopoietic stem cells do not transdifferentiate into cardiac myocytes in myocardial infarcts. Nature 428:664–668
Orlic D, Kajstura J, Chimenti S et al 2001 Bone marrow cells regenerate infarcted myocardium. Nature 410:701–705
Park K-S, Lee D-k, Lee H et al 2003 Phenotypic alteration of eukaryotic cells using randomized libraries of artificial transcription factors. Nat Biotechnol 21:1208–1214
Ramalho-Santos M, Yoon S, Matsuzaki Y, Mulligan RC, Melton DA 2002 'Stemness': transcriptional profiling of embryonic and adult stem cells. Science 298:597–600
Richards M, Fong C-Y, Chan W-K, Wong P-C, Bongso A 2002 Human feeders support prolonged undifferentiated growth of human inner cell masses and embryonic stem cells. Nat Biotechnol 20:933–936
Rideout WM, III, Hochedlinger K, Kyba M, Daley GQ, Jaenisch R 2002 Correction of a genetic defect by nuclear transplantation and combined cell and gene therapy. Cell 109:17–27
Sato N, Sanjuan IM, Heke M, Uchida M, Naef F, Brivanlou AH 2003 Molecular signature of human embryonic stem cells and its comparison with the mouse. Dev Biol 260:404–413
Sato N, Meijer L, Skaltsounis L, Greengard P, Brivanlou AH 2004 Maintenance of pluripotency in human and mouse embryonic stem cells through activation of Wnt signaling by a pharmacological GSK-3-specific inhibitor. Nat Med 10:55–63
Schuldiner M, Yanuka O, Itskovitz-Eldor J, Melton DA, Benvenisty N 2000 Effects of eight growth factors on the differentiation of cells derived from human embryonic stem cells. Proc Natl Acad Sci USA 97:11307–11312
Shamblott MJ, Axelman J, Wang S et al 1998 Derivation of pluripotent stem cells from cultured human primordial germ cells. Proc Natl Acad Sci USA 95:13726–13731 (Erratum in Proc Natl Acad Sci USA 96:1162)
Solter D 1999 Cloning and embryonic stem cells: a new era in human biology and medicine. Croat Med J 40:309–318
Solter D, Gearhart J 1999 Putting stem cells to work. Science 283:1468–1470
Solter D, Beyleveld D, Friele MB et al 2003 Embryo research in pluralistic Europe. Springer-Verlag, Berlin
Sperger JM, Chen X, Draper JS et al 2003 Gene expression patterns in human embryonic stem cells and human pluripotent germ cell tumors. Proc Natl Acad Sci USA 100:13350–13355
Thomson JA, Itskovitz-Eldor J, Shapiro SS et al 1998 Embryonic stem cell lines derived from human blastocysts. Science 282:1145–1147

Wagers AJ, Weissman IL 2004 Plasticity of adult stem cells. Cell 116:639–648
Wagers AJ, Sherwood RI, Christensen JL, Weissman IL 2002 Little evidence for developmental plasticity of adult hematopoietic stem cells. Science 297:2256–2259
Wakayama T, Tabar V, Rodriguez I, Perry ACF, Studer L, Mombaerts P 2001 Differentiation of embryonic stem cell lines generated from adult somatic cells by nuclear transfer. Science 292:740–743
Weimann JM, Johansson CB, Trejo A, Blau HM 2003 Stable reprogrammed heterokaryons form spontaneously in Purkinje neurons after bone marrow transplant. Nat Cell Biol 5:959–966
Weissman IL 2000 Stem cells: units of development, units of regeneration, and units in evolution. Cell 100:157–168
Wichterle H, Lieberam I, Porter JA, Jessell TM 2002 Directed differentiation of embryonic stem cells into motor neurons. Cell 110:385–397
Ying Q-L, Nichols J, Chambers I, Smith A 2003 BMP induction of Id proteins suppresses differentiation and sustains embryonic stem cell self-renewal in collaboration with STAT3. Cell 115:281–292

DISCUSSION

Gearhart: David Prentice, a professor at Indiana State University, has compiled lists of publications purported to demonstrate that adult tissues are an effective source of stem cells and argues that the studies on embryonic sources of stem cells are not needed. He frequently testifies before Congress and various committees, presenting long lists of references demonstrating the efficacy of adult stem cells (See, for example, Appendix K, with its 192 references, in *Monitoring Stem Cell Research, A Report of the President's Council on Bioethics*, January, 2004). Irv Weissman and others have gone through the long process of reading all these references and have come to the conclusion that Prentice's list is misleading and his claims overstated.

Solter: The problem is that one can say that what each of these papers presents is not truly proven, but one cannot say it is obviously not true. This is the difficulty. Much of this research has been done in ways that are not as rigorous as we would think necessary.

Weissman: I went through all of these papers to see whether they matched the following criteria. (1) That they started with a defined adult stem cell of one type, marked or purified. (2) That there was clear transdifferentiation or that another cell type came up. (3) That it was human. None of the papers met all three criteria.

Björklund: Would it make a difference if you excluded the third criterion?

Weissman: It made little difference. There were just a few papers. I informed Prentice and his colleagues about this a long time ago, but they keep saying the same thing.

Grompe: This is not a trivial issue in the sense that one would have thought this would have been resolved in the scientific community. Yet if one looks at reviews on this topic published in 2003 there are some clear statements, such as the following: 'it is well accepted that plasticity and fusion both exist and are equally important'. This is the prevailing attitude in the scientific community.

Weissman: Not in the cognoscenti!

Grompe: That's what everyone always thinks about their own opinion.

Weissman: There is no clear evidence that one approach is superior to the other at this point in time. The argument should not only be based around therapeutic cloning: taking a cell from a person, making a cell line and then treating that person. This is laudable and I hope we do it. But the argument for nuclear transfer to produce human pluripotent stem cell lines is that you want to have cell lines from predefined donors who have genetically determined diseases, so that you have a chance to understand the pathogenesis of the diseases. This seems to be a political and religious opinion to ban a line of research. If we take the point of view that therapeutic transplantation is the only use of cloning, you will fall into their trap and it will be damaging to us as scientists.

Gearhart: Those opposed to therapeutic cloning argue that it hasn't even cured any mouse diseases. Marcus Grompe, are you saying that ES cells haven't been shown in principle to work therapeutically?

Grompe: Which of the ES therapeutic papers are convincing to you?

Weissman: Brustle et al (1999) on oligodendrocytes, and George Dailey with Rudi Jaenisch (Rideout et al 2002) showing that it might at least be possible to create ES cells derived from haematopoietic stem cells that, with some modification, grow up to be useful.

Grompe: I can't say that I have read all these papers, but I tend to agree with Irv Weissman. There has been no convincing evidence of true transdifferentiation or plasticity. In the area that I work on, the liver, I can say with certainty that I haven't seen a single paper that convinces me that there is actual plasticity of any cell type. Those papers that talk about plasticity and which claim significant tissue replacement, in the order of 1–10% in the liver, are truly sloppily done, and these results have not held up when people have tried to reproduce them. Those events that are potentially consistent with plasticity are extremely rare. I don't know of a single example where there is physiologically relevant plasticity in the sense that a haematopoietic stem cell or a neural stem cell could actually fix a physiological problem: liver disease or Parkinson's, for example. If plasticity or transdifferation truly occurs, at the current time it is not by a physiological mechanism such as some recent reviews suggest which describe circulating cells that could go around and cure every disease on earth. If it truly exists it may be a process that could be harnessed therapeutically, but currently it seems to be a rare, physiologically unimportant mechanism.

Gurdon: As I understand it, the problem is defining any state of the starting cell population in these cases that you mention. You find cells that differentiate in various directions, but the problem is that we don't know what kind of cell gives rise to the cells that subsequently have a clearly defined phenotype.

Grompe: I don't think that is necessarily the problem. There is one very strong paper by Diane Krause and Saul Sharkis (Krause et al 2001) in which they used a cell that is as precisely defined as possible. This is a single haemopoietic stem cell (HSC). In this paper they claim that they get 20% of the pulmonary epithelium from a single HSC. This is something that others simply have not been able to reproduce. The cell input in many cases is quite well defined; what has not been done is that in very few papers have people looked for genetic material from the host. Another problem is that people simply use cell surface markers or immunohistochemistry as their definition of transdifferentiation, and not functional assays that look at a whole range of properties. If you get a cell that is cytokeratin positive, does that mean it is a true pulmonary epithelial cell? I would say that is not enough. There are two problems: firstly, the cell that the starting cell differentiates into is poorly defined, and secondly, host genetic markers have not been looked at systematically.

Gurdon: In many of the papers that you refer to, presumably the problem is that it isn't possible to define definitively the differentiated state of one cell out of a population when you see differentiated cells of different kinds appearing. I don't understand how, if you inject a million cells, you can be sure which was the precursor to the cells that are subsequently seen.

Grompe: That is true, but there are examples in that list of papers where people have used clearly defined individual cells, retrovirally marked and so forth. In many papers it is not the fact that the input population was poorly defined: the problem is with characterizing the output population.

Nadal-Ginard: Are you saying that in all these papers claiming differentiated phenotypes or cells coming from the bone marrow, that none of these differentiated phenotypes are true differentiated phenotype cells coming from bone marrow?

Grompe: In terms of having truly functional properties, there are very few papers. There are a few recent examples. The paper by Helen Blau showing Purkinje neurons coming from bone marrow is convincing (Weimann et al 2003). A Purkinje neuron is morphologically well defined. I have not seen any papers showing the transition from bone marrow to cardiomyocyte from HSCs.

Nadal-Ginard: I didn't say they were from the HSCs, but rather from the bone marrow. Why do they have to be from HSCs to be valid?

Grompe: This is necessary for it to be true plasticity. My definition of stem cell plasticity is when a stem cell crosses the germline lineage boundaries. I have no trouble with the idea that mesenchymal stem cells can take on muscle lineages. This is not quite the same phenomenon as plasticity.

Nadal-Ginard: This depends on the definition of 'plasticity'. Crossing germ layers is one type of plasticity. Going from one tissue type to another is another type.

Solter: There are stromal cells in the heart, essentially fibroblasts. These can become adipocytes or muscle cells. This means that there is a stem cell precursor of several mesenchymal-type tissues in the heart, which could explain plasticity.

Nadal-Ginard: We are confusing our arguments here. On the one hand we are looking for cells that are useful as precursors of tissues. On the other hand, when we have these cells we say that this is not enough, because we don't know their precise origin, and this is not considered to be plasticity. If we are talking about reconstituting the tissue, then this is a valid approach no matter where the cell comes from. The arguments need to be separated.

Solter: What is crucial here is that there is an attempt by some to say that ES cells should never be used because we can prove that adult stem cells can turn into any cell we want. This is what we are disputing.

Trounson: I'd like to ask you about the definitions: these may be the cause of some of the problems we are having. What about pluripotentiality relating to tumour formation and the type of cells you can form when you transplant these cells *in vivo*? In the mouse you can clearly transplant the cells into an embryo and see whether they contribute to all the cells in the embryo, but do any of these cells except for ES, EC and MAPs really make a genuine teratoma when they are placed *in vivo*? I think you suggested that placental cells might. Should the ability to form teratomas be used to create a better definition of pluripotentiality?

Solter: Do you mean that a truly pluripotential cell is one that produces a complex teratoma?

Trounson: Yes

Solter: I suppose so. This is more-or-less the only way we can test for the pluripotency of human stem cells.

Soria: We could talk about functional differentiation and not only tumours. Some precursors could be put into a SCID mouse and then develop into a different cell type. It would be nice if we could reach some conclusions about the criteria for differentiation. Marcus Grompe raised the issue of functional reconstitution: this is one way to test whether the cell has a certain degree of differentiation because the list of markers is never complete. This could be one criterion.

Trounson: These kinds of things are important. I frequently read papers where people have used markers such as *Oct4* to indicate a pluripotential cell. I am not sure that this fits all our required definitions of functionality, tumour formation and so on. There are cells that express genes or produce markers that don't fit the definition of pluripotentiality that we are talking about. In the area of transdifferentiation we have to be careful about what we are defining as the cell we are focusing on. Mesenschymal stem cells, for instance, seem to be a little different than HSCs or other adult stem cells in their ability to form other tissues. There are perhaps further definitions of multipotentiality that we haven't determined as yet.

Gurdon: What would you regard as a test of true function?

Trounson: In vivo or *in vitro*? *In vivo* you really have to demonstrate a nerve cell that is able to pass a message or have a characteristic that is commonly attributed to those cells. A cardiomyocyte would need to have the electrophysiological characteristics you would expect. I think for specific function you can ask specialists to provide the primary definition of function for those cells. If you can create tests of these functions *in vitro*, this would be helpful but sometimes they will only function when they are integrated with other cells in some kind of organization.

Gurdon: Then it is difficult to decide that the function seen is attributable to the particular cell rather than to the others around it.

Trounson: If they can't function with the other cells around them one might be concerned that they are resident cells without genuine function.

Grompe: It is important to remember that part of the reason that the HSC field is so far ahead of all the other somatic stem cells is the fact that the reconstitution assay is available and therefore people have been able to stringently define what an HSC truly is. It is one that when transplanted can rescue lethally radiated animals and establish all the haematopoietic lineages *in vivo*. For many of the other stem cell systems, similar assays do not exist.

Wilmut: The haematopoietic system is very different in that there is a very rapid turnover. Is it likely that a single stem cell from other tissues would have the ability to restore physiological function?

Nadal-Ginard: In solid organs the one cell assay is not available. But this is the criterion that one has to meet in order to convince the sceptic. How is it going to be met?

Solter: Would it be possible to make a parabiotic animal and do complete hepatectomy in one, and then try to reconstitute the liver in that animal by providing a single stem cell?

Nadal-Ginard: Then you are putting two hurdles in front of the cells. One is being able to differentiate, the other is being able to organize not only a tissue but also an organ.

Gepstein: There are several tissues for which you could isolate the cells after they have transdifferentiated or not, and try to prove the functionality. In the heart we can isolate our transplanted cells and look at their electrophysiological properties. We can prove functionality, which in my view is much more important than just showing one or two markers.

Wilkins: We have been discussing operational criteria, but I would like to ask a basic question about definitions. I think they are important because they allow us to decide whether we are talking about the same entity. Is it time to throw out the textbook definition of a stem cell as one that when it divides gives rise to one daughter cell with a wide range of potentialities and another which is dedicated to making a lineage of some specialized cell type? Is there evidence that this definition is no longer any good, and if we reject it are we reducing the term stem cell to mean simply some kind of pluripotent or multipotent cell?

Solter: Certainly, the stem cell for many divisions makes nothing but stem cells. In development there are certainly situations in which we can prove that one cell gives two different daughters. Previously people argued that there is quantal mitosis: the stem cell divides a number of times making just stem cells, and then after a given number of divisions the stem cells divide into two cells which are identical to each other but different from the cell they came from. These are differentiated cells. I don't know whether this happens.

Wilkins: So are you saying that the textbook definition should now be thrown out, and the term 'stem cell' will just become a synonym for pluripotent or multipotent cells?

Solter: I guess so.

Nadal-Ginard: Do stem cells age and become senescent?

Solter: For ES cells we claim that they can live forever, but I don't know how we can be sure of this.

Nadal-Ginard: We select the ones that survive.

Björklund: Markus Grompe talked about the difficulty of being sure of differentiation of stem cells *in vivo*. In the brain this is partly a technical problem. Cells from the bone marrow readily enter the brain in some areas. When using labels or markers that are confined to the cell's nucleus there is a real problem that small non-neuronal cells, including microglia and cells that remain undifferentiated, are closely attached to the surface of neurons. In identifying the labelled nucleus it is critical to do the identification cell by cell. One can safely conclude that those papers that have reported widespread neuronal differential from blood-derived cells have not performed careful identification of the cells. The only convincing case is Purkinje cells in the cerebellum. This seems, however, to be due to cell fusion, occurs in a very low numbers, and takes a long time.

Goldman: This gets back to the issue of rigorously proving the differentiated phenotype of the daughter cells. In the nervous system I am not aware of any example of a non-neural cell line giving rise to neurons where it is has been proven that the criteria for differentiated functional neurons, as classically defined by action potentials, synaptic integration and synaptic function, have been fulfilled except for neurons derived from neural stem cells or ES cells.

Murray: I want to underscore the difficulty of translating the sophistication of this conversation into the policy arena. Prentice's name has been mentioned. What has happened is a replication of a strategy that was hit on at least 30 years ago by the asbestos industry, in claiming that the health consequences of asbestos were a 'scientific controversy'. All one needed was a handful of scientists to say that it was not proven. You could have 10 000 on the other side, but if you could bring two or three to a hearing you could claim that it was a controversy. The asbestos industry stumbled onto this strategy; the tobacco industry refined and perfected it. They ended up with a fairly sophisticated procedure for recruiting scientists: they needed a steady supply because eventually each scientist would erode his or her credibility. They would have middle men to call scientists and try to vet them, to see whether they were at least moderately sympathetic to the industry's position. If so the industry would exalt them, publish them in symposia and cite their data. One sees this pattern skilfully duplicated in the arena of adult stem cells.

Minger: I am an American who has lived in the UK for 8 years. The problem with this issue is the politicization: it is not a basic biology problem. In the UK it is not that big an issue: whether transdifferentiation exists is not a political issue but a biological one. In this country you can get a licence to generate human ES cells and it is legal to work on them.

Murray: The UK is an exception rather than the rule, even within Europe, in terms of the clarity of its position.

Minger: The Danes and Swedes have passed legislation that is very similar to the UK.

Gearhart: The more we learn about human ES cells compared with the mouse ES cells, the more we find they are different. When we look at some of the gene profiles and differentiating protocols, there are significant differences. We routinely find trophoblast differentiating in cultures of human ES but not with mouse ES. What is the basis of this? My concern is that everyone views these mouse ES and human ES cells as being identical.

Lovell-Badge: My view is that there is no real natural homologue in the embryo to an ES cell, either in mouse or human. What you are doing is providing an *in vitro* niche which allows you to stabilize some cell type. There is no reason why these should be identical. This is borne out by the data. There will be things in common, and I think it is what they do next that is important. They will follow similar cues that occur in the embryo in terms of their ability to differentiate. You can get the same things from them, but it doesn't surprise me at all if they are different.

Solter: It is true that they are different: as you point out, human ES cells can turn into trophoblast cells but mouse ES cells cannot. This might mean that human ES cells were side-tracked from the embryo a little bit earlier, or that the capacity of human inner cell mass cells to generate trophoblast persists longer. It is true that ES cells are *in vitro* culture artefacts. We took an embryo and then somehow managed to preserve a cell for ever which in the embryo exists for perhaps a day or two, whereupon they all differentiate and disappear for ever. The other thing is that they still might be quite similar. I would not be surprised if someone figured out a way to turn mouse ES cells into trophoblast by skilfully applying some growth factor. Each mammal seems to be different. There must be a reason why we are unable to derive ES cells from other mammals except for mice and perhaps humans. A major effort has been put into making farm animal and rat ES cells but this has so far failed.

Lovell-Badge: We can turn mouse ES cells into trophoblast by down-regulating *Oct4* or, at least with inner cell mass cells, by inactivating *Sox2*.

Trounson: There are also some strange differences within the ES cell lines. We are unable to differentiate some of our human ES cell lines into certain lineages if they have gone beyond a certain passage. If you are comparing the cells pre and post that passage they would have different characteristics. I am not surprised you see vast differences on microarrays. I suspect there are differences among all the ES cell lines that have been raised, and you could probably show quite a lot of differences between them in both their ability to form other tissues or the ease with which they form other tissues under some drivers of differentiation. We haven't recognized these important issues yet. It is a little bit like the mouse, in a way. We ended up working on very few mouse ES cell lines because they worked well across many laboratories. I wonder whether we might see something like that in the human: we might end up selecting one line out of 300 which is useful experimentally but which might be quite different to others.

References

Brustle O, Jones KN, Learish RD et al 1999 Embryonic stem cell-derived glial precursors: a source of myelinating transplants. Science 285:754–756

Krause DS, Theise ND, Collector MI et al 2001 Multi-organ, multi-lineage engraftment by a single bone marrow-derived stem cell. Cell 105:369–377

Rideout WM 3rd, Hochedlinger K, Kyba M, Daley GQ, Jaenisch R 2002 Correction of a genetic defect by nuclear transplantation and combined cell and gene therapy. Cell 109:17–27

Weimann JM, Johansson CB, Trejo A, Blau HM 2003 Stable reprogrammed heterokaryons form spontaneously in Purkinje neurons after bone marrow transplant. Nat Cell Biol 5: 959–966

Bone marrow-derived hepatocytes

Markus Grompe

Department of Molecular and Medical Genetics L103, Oregon Health & Science University, 3181 SW Sam Jackson Park Road, Portland, OR 97239, USA

Abstract. Several independent reports have demonstrated that bone marrow cells can give rise to different hepatic epithelial cells types, including oval cells, hepatocytes and duct epithelium. These observations have resulted in the hypothesis that bone marrow resident stem cells, specifically haematopoietic stem cells, are an important source for liver epithelial cell replacement, particularly during chronic injury. However, the bulk of published data on the role of bone marrow stem cells in liver damage suggest that they do not play a significant physiological role in replacement of epithelial cells in any known form of hepatic injury. Fully functional bone marrow-derived hepatocytes indeed exist, but are extremely rare and are generated by cell fusion, not stem cell differentiation.

2005 Stem cells: nuclear reprogramming and therapeutic applications. Wiley, Chichester (Novartis Foundation Symposium 265) p 20–34

Oval cells and hepatocytes after bone marrow transplantation

Bone marrow transplantation has been used experimentally and in clinical medicine for several decades to reconstitute the haematopoietic system, including the immune system of hosts with donor-derived cells (Thomas et al 1975a,b). In the late 1990s several reports on unexpected plasticity of bone marrow-derived stem cells were published (Bjornson et al 1999, Ferrari et al 1998, Mezey et al 2000). These reports on stem cell plasticity and the observed expression of haematopoietic markers in hepatic oval cells led to the hypothesis that bone marrow stem cells may also give rise to epithelial cells including hepatic oval cells. This was confirmed experimentally in 1999 by Petersen et al (1999). Female rats were transplanted with male bone marrow cells and exposed to the classic AAP/partial hepatectomy protocol which results in oval cells expressing c-kit (Omori et al 1997). Although not reported in the published paper, the authors also performed the reverse experiment in which orthotopic liver transplantation from female into male donors was performed (B. Petersen, personal communication). In both cases, some of the emerging oval cells contained the donor Y-chromosome marker. This seminal paper therefore provided proof-of-principle that liver epithelial cells could indeed be bone marrow-derived.

Other reports soon confirmed the basic observation and added new information. A murine study showed that not only oval cells, but also hepatocytes could be derived from donor bone marrow (Theise et al 2000a). Human patients which had undergone a gender-mismatched bone marrow or liver transplantation were also examined using fluorescent *in situ* hybridization (FISH) as the main method to detect bone marrow-derived epithelial cells. Two groups independently reported their observation that bone marrow-derived hepatocytes could be frequently found in such patients (Theise et al 2000b, Alison et al 2000). Remarkably one patient was reported to have as many as 30% of hepatocytes of bone marrow origin (Theise et al 2000b). Subsequent papers also reported human epithelial cells in bone marrow transplantation (BMT) patients (Korbling et al 2002).

Although the basic observation of bone marrow-derived hepatic epithelial cells was therefore firmly established, two important questions remained. First, it was unknown which cell type in the marrow was the hepatocyte precursor. Second, the functionality of these cells was uncertain. Lagasse et al (2000) clarified this issue by showing that the bone marrow hepatocyte precursor co-purified with the haematopoietic stem cell (HSC) in cell sorting experiments. Only cells which were c-kit^{+}, Lin^{-}, Sca-1^{+} and Thy-1^{lo} (KLTS cells) gave rise to hepatocytes. This result suggested, but did not definitely prove that the HSC itself was the hepatocyte precursor. Because the lowest cell dose of KLTS cells used was 50, heterogeneity within this population still needed to be considered. This study also demonstrated extensive liver repopulation and a complete correction of the metabolic liver disease, hereditary tyrosinemia (Lagasse et al 2000), thus showing conclusively that bone marrow-derived hepatocytes function normally. The metabolic correction involved multiple hepatocyte specific parameters such as plasma amino acid levels, bilirubin conjugation and excretion and serum transaminase levels.

By performing single cell transplants Krause and co-workers subsequently showed that a single HSC could reconstitute not only the haematopoietic system of recipient mice but also contribute to multiple other tissues including liver epithelium (Krause et al 2001). Interestingly, in this experiment, donor cell markers were found only in rare hepatic ducts but not hepatocytes. It should also be noted that the bone marrow stem cell type used was not isolated using the KLTS phenotype but instead used a quite different protocol (Lanzkron et al 1999). Therefore it is currently not known whether the single cell reconstitution with a KLTS stem cell would produce similar results.

Physiological significance of bone marrow-derived epithelial cells

Despite the unambiguous finding that fully functional bone marrow-derived hepatocytes really exist, there has been considerable controversy regarding their functional importance in liver injury. In one model, replenishment of liver cells from the bone marrow is an important injury response pathway, particularly in progenitor dependent regeneration (oval cell response). Hepatocytes would be generated either by direct differentiation of HSCs or indirectly via an oval cell intermediate. The opposing model suggests that oval cell precursors reside only in the liver itself (tissue resident stem cell) and that the bone marrow contributes very few liver epithelial cells, even during injury. Some of the early reports on the incidence of bone marrow-derived hepatocytes suggested by high levels of replacement (1–2%) approaching the therapeutic threshold for some genetic liver disorders (Theise et al 2000a). Such a high replacement level only two months after BMT and without any liver injury other than whole body irradiation suggested that the bone marrow served as a very important source of hepatocytes. This was surprising because even intrahepatic transplantation with single cell suspensions of differentiated hepatocytes does not result in a higher degree of repopulation (Ponder et al 1991). The quantitative results of the Theise study (Theise et al 2000a) suggested that BMT should be considered as a therapeutic option for the treatment of liver diseases. Unfortunately, however, multiple subsequent studies in bone marrow-transplanted rodents independently failed to confirm the initial report (Gao et al 2001, Mallet et al 2002, Wagers et al 2002). In all of these papers, the observed degree of hepatocyte/epithelial cell replacement after BMT was at least three orders of magnitude lower than 1% and was 1/100 000 or less even many months after BMT. These findings raise the issue whether the use of Y-chromosome FISH as a donor marker may markedly overestimate the frequency of hepatocyte replacement (Wagers et al 2002). After all, all of the tissue resident donor-derived haematopoietic cells (macrophages, lymphocytes, etc.) also contain a Y-chromosome. Furthermore, it should be noted that optimal fixation/tissue preparation methods for *in situ* hybridization and immunohistochemistry are quite different. For this reason, co-localization studies showing the expression of epithelial markers and the presence of the Y-chromosome marker in the same cell are difficult and prone to artefact, particularly in archival specimens such as those frequently used for human studies. The highest levels of epithelial cell replacement after BMT have been reported in retrospective histopathological specimens from human patients (Theise et al 2000b, Alison et al 2000). The marked discrepancy to multiple rodent studies could be due to:

- positive selection of bone marrow-derived hepatocytes in some human patients
- technical difficulties in the co-localization or
- an important physiological role for human bone marrow cells in liver replacement.

Positive selection is a distinct possibility in these human cases, because most patients had received chemotherapy for haematological malignancies before transplantation and/or had hepatocyte injury from viral hepatitis.

To determine whether liver damage enhances the transition of HSCs to hepatocytes, the degree of cell replacement after BMT in mice has been measured not only in healthy animals, but also in the context of pre-existing hepatocyte injury (Wang et al 2002) or oval cell regeneration (Wang et al 2003a). Importantly, the frequency of bone marrow-derived hepatocytes was not higher in the acute liver injury seen in hereditary tyrosinemia type 1 than in healthy control livers (Wang et al 2002). In addition, no contribution from bone marrow precursors to the oval cell reaction induced by the chemical 3,5-diethoxycarbonyl-1,4-dihydrocollidine was detected (Preisegger et al 1999, Wang et al 2003a). Together these results indicate that HSCs definitely do not serve as epithelial cell precursors in all forms of hepatic injury. Therefore, the presence of haematopoietic markers on oval cells does not necessarily signify a lineage relationship and an origin in the bone marrow. It remains to be determined whether HSCs play a significant role in any clinically relevant hepatic injury models to be examined in the future.

Mechanism of the conversion of bone marrow stem cells to hepatocytes

Three basic mechanisms for the emergence of donor hepatocytes after BMT can be considered. First, bone marrow could theoretically harbour a specialized endodermal stem cell capable of producing hepatocytes and other epithelial cells. This cell would be analogous to the mesenchymal stem cell which produces mesodermal derivatives such as muscle, cartilage and fat (Pereira et al 1995, Prockop 1997). Second, hepatocytes and blood cells could be derived from the same stem cell by hierarchical differentiation. Until recently, this has been the hypothesis favoured by many in the field (Petersen 2001). Third, bone marrow-derived hepatocytes could potentially be derived not from differentiation at all, but rather by cell fusion. This possibility was raised by the observation that haematopoietic cells could spontaneously fuse with embryonic stem cell *in vitro* and then produce multiple tissues in chimeric mouse embryos (Ying et al 2002, Terada et al 2002). To date, most experiments in the stem cell plasticity field were not designed to discern whether fusion could represent a potential mechanism for apparent plasticity. Positive markers such as the Y-chromosome, green fluorescent protein or β-galactosidase were only present in the donor cells and it was therefore not possible to determine whether the cells containing donor markers also contained host genes. In order to address this question in the only animal model in which bone marrow had functionally reconstituted injured liver, the tyrosinemia mouse model, transplantation experiments were designed in which genetic markers for both donor marrow and host hepatocytes were used. Bone marrow-derived hepatocytes were serially transplanted in order to avoid confusion by haematopoietic cell contamination in repopulated liver (Wang et al 2003b). The results from this study were unambiguous: the vast majority of bone marrow-derived hepatocytes contained genetic information from both the donor and host, indicating cell fusion. Cytogenetic analysis of female→male gender mismatched transplants indicated a high frequency of tetraploid XXXY and hexaploid XXXXYY karyotypes, predicted for fusion. Importantly, similar experiments were independently performed by another laboratory and confirmed that fusion was the predominant mechanism for derivation of hepatocytes from bone marrow stem cells in the tyrosinemia model (Vassilopoulos et al 2003, Wang et al 2003b). Analysis of retroviral integration sites in a single animal also suggested a clonal origin for hepatocytes and blood cells, the expected result for cell fusion.

The clear evidence for cell fusion in bone marrow-derived hepatocytes raises the issue whether fusion is a primary or secondary event. Do HSCs differentiate into bona fide hepatocytes first and then secondarily fuse with pre-existing hepatocytes? Multiple lines of evidence suggest that fusion is the primary event, not secondary. First, extensive literature exists on the phenomenon of polyploidy in hepatocytes. The bulk of these reports indicates failed cytokinesis, not fusion, is responsible for the presence of binucleated and polyploid hepatocytes (Yim 1982, Auvigne et al 2002, Faktor & Uryvaeva 1975). Second, no evidence of cell fusion was detected in *Fah* mutant control animals which had undergone several rounds of liver damage and regeneration (Wang et al 2003b). Hexaploid karyotypes were found exclusively in bone marrow-derived hepatocytes, never in wild-type or *Fah* mutant controls.

The detailed cytogenetic analysis of serially transplanted bone marrow-derived hepatocytes revealed that some of the fusion hepatocytes were diploid (Wang et al 2003b). This was an unexpected finding and suggested that the fusion cells could undergo a reduction division thereby reducing their chromosome number (Martin & Sprague 1970). The fact that the host Y-chromosome and the donor selectable marker *Fah* were present in the same cell indicates that the initially binucleated heterokaryons went through a mononucleated stage followed by random segregation of the chromosomes in a subsequent mitosis. This finding has significant implications for the debate on stem cell plasticity, because showing diploidy in donor-derived cells is clearly not sufficient to rule out cell fusion as the mechanism. To rule out fusion, independent positive markers for both the donor and host genome must be assessed. None of the positively identified donor-derived cells should have the host marker.

Cytogenetic analysis also revealed the existence of many fusion hepatocytes with numerous chromosome aberrations (aneuploidy) (Wang et al 2003b). It is currently not clear whether this is the result of abnormal chromosome segregation during cell division or is caused by pre-existing chromosome abnormalities in *Fah* mutant hepatocytes. Importantly, no tumours have been found to originate from bone marrow-derived hepatocytes even when they were serially transplanted and underwent many cell divisions. Therefore, the observed aneuploidy does not seem to represent a precancerous event.

Myelomonocytic cells are the source of bone marrow-derived hepatocytes

Although the majority of bone marrow-derived hepatocytes clearly emerge from fusion between donor bone marrow-derived cells and host hepatocytes, the exact nature of the haematopoietic cells fusing with hepatocytes was not known. In principle any of the nucleated blood cells derived from the HSCs could be considered potential fusion partners. The HSC itself or any of its differentiated progeny could be the fusion donor. For this reason our laboratory recently performed cell fractionation and genetic experiments to determine whether all blood cells or only subpopulations can produce bone marrow-derived hepatocytes (Willenbring et al 2004). First, pure HSCs from lymphocyte-deficient $Rag1^{-/-}$ mice were transplanted into *Fah* null mice. Despite the complete absence of B or T cells, bone marrow-derived hepatocytes were produced at robust levels in these experiments. This indicated that lymphocytes are not required for the generation of bone marrow-derived hepatocytes and pointed toward myelomonocytic cells as their likely source. This was confirmed by using as donors lineage-committed granulocyte/macrophage progenitors (GMPs) sorted to purity using cell surface markers. Despite the complete absence of long-term engrafting HSCs in the donor population, many clusters of bone marrow-derived hepatocytes were generated in multiple independent experiments. Therefore, cells of the myelomonocytic lineage are sufficient for fusion and HSCs themselves are not needed. This is one of the strongest arguments against stem cell plasticity as the source of bone marrow-derived epithelial cells *in vivo*. These experiments were taken a step further by using terminally differentiated bone marrow-derived macrophages (BMMs) as transplantation donors (Willenbring et al 2004). The recipient *Fah* mutant did not receive any preparative radiation and there was no evidence for long-term haematopoietic engraftment. Nonetheless, bone marrow-derived hepatocytes were once again produced at robust levels. These results—recently also confirmed by others (Camargo et al 2004)—provide direct evidence that committed myelomonocytic cells such as macrophages can produce functional epithelial cells by *in vivo* fusion.

Therapeutic use of bone marrow-derived hepatocytes

The complete metabolic correction achieved by bone marrow-derived hepatocytes in the murine model of hereditary tyrosinemia type 1 illustrates that these cells, even if generated by cell fusion rather than stem cell differentiation, are fully functional and hence therapeutically useful. However, spontaneous cell fusion is a very rare event even in the presence of significant liver injury and therefore therapeutic effects can be achieved only by growth selection of the very few fusion hepatocytes (Mitchell & Fausto 2002). Strategies to create an *in vivo* selection of the donor cells were the obvious solution to achieve therapeutic benefit using bone marrow as a source for hepatocytes. Now, the discovery of cell fusion as the underlying mechanistic principle suggests a second approach to improve efficiency: induced fusion between host hepatocytes and transplanted myelomonocytic donor cells. Many chemicals and proteins are known to induce fusion between adjacent cells and are now candidates to enhance the usefulness of macrophage transplantation in liver disease.

References

Alison MR, Poulsom R, Jeffery R et al 2000 Hepatocytes from non-hepatic adult stem cells. Nature 406:257

Auvigne I, Pichard V, Aubert D, Robillard N, Ferry N 2002 In vivo cell lineage analysis in cyproterone acetate-treated rat liver using genetic labeling of hepatocytes. Hepatology 35:281–258

Bjornson CR, Rietze RL, Reynolds BA, Magli MC, Vescovi AL 1999 Turning brain into blood: a hematopoietic fate adopted by adult neural stem cells in vivo. Science 283:534–537

Camargo FD, Finegold M, Goodell MA 2004 Hematopoietic myelomonocytic cells are the major source of hepatocyte fusion partners. J Clin Invest 113:1266–1270

Faktor VM, Uryvaeva IV 1975 Progressive polyploidy in mouse liver following repeated hepatectomy. Tsitologiia 17:909–916

Ferrari G, Cusella-De Angelis G, Coletta M et al 1998 Muscle regeneration by bone marrow-derived myogenic progenitors. Science 279:1528–1530

Gao Z, McAlister VC, Williams GM 2001 Repopulation of liver endothelium by bone-marrow-derived cells. Lancet 357:932–933

Korbling M, Katz RL, Khanna A et al 2002 Hepatocytes and epithelial cells of donor origin in recipients of peripheral-blood stem cells. N Engl J Med 346:738–746

Krause DS, Theise ND, Collector MI et al 2001 Multi-organ, multi-lineage engraftment by a single bone marrow-derived stem cell. Cell 105:369–377

Lagasse E, Connors H, Al-Dhalimy M et al 2000 Purified hematopoietic stem cells can differentiate into hepatocytes in vivo. Nat Med 6:1229–1234

Lanzkron SM, Collector MI, Sharkis SJ 1999 Hematopoietic stem cell tracking in vivo: a comparison of short-term and long-term repopulating cells. Blood 93:1916–1921

Mallet VO, Mitchell C, Mezey E et al 2002 Bone marrow transplantation in mice leads to a minor population of hepatocytes that can be selectively amplified in vivo. Hepatology 35:799–804

Martin GM, Sprague CA 1970 Vinblastine induces multipolar mitoses in tetraploid human cells. Exp Cell Res 63:466–467

Mezey E, Chandross KJ, Harta G, Maki RA, McKercher SR 2000 Turning blood into brain: cells bearing neuronal antigens generated in vivo from bone marrow. Science 290:1779–1782

Mitchell C, Fausto N 2002 Bone marrow-derived hepatocytes : rare but promising. Am J Pathol 161:349–350

Omori M, Evarts RP, Omori N et al 1997 Expression of alpha-fetoprotein and stem cell factor/c-kit system in bile duct ligated young rats. Hepatology 25:1115–1122

Pereira RF, Halford KW, O'Hara MD et al 1995 Cultured adherent cells from marrow can serve as long-lasting precursor cells for bone, cartilage, and lung in irradiated mice. Proc Natl Acad Sci USA 92:4857–4861

Petersen BE 2001 Hepatic 'stem' cells: coming full circle. Blood Cells Mol Dis 27:590–600

Petersen BE, Bowen WC, Patrene KD et al 1999 Bone marrow as a potential source of hepatic oval cells. Science 284:1168–1170

Ponder KP, Gupta S, Leland F et al 1991 Mouse hepatocytes migrate to liver parenchyma and function indefinitely after intrasplenic transplantation. Proc Natl Acad Sci USA 88:1217–1221

Preisegger KH, Factor VM, Fuchsbichler A et al 1999 Atypical ductular proliferation and its inhibition by transforming growth factor beta1 in the 3,5-diethoxycarbonyl-1,4-dihydrocollidine mouse model for chronic alcoholic liver disease. Lab Invest 79:103–109

Prockop DJ 1997 Marrow stromal cells as stem cells for nonhematopoietic tissues. Science 276:71–74

Terada N, Hamazaki T, Oka M et al 2002 Bone marrow cells adopt the phenotype of other cells by spontaneous cell fusion. Nature 416:542–545

Theise ND, Badve S, Saxena R et al 2000a Derivation of hepatocytes from bone marrow cells in mice after radiation-induced myeloablation. Hepatology 31:235–240

Theise ND, Nimmakayalu M, Gardner R et al 2000b Liver from bone marrow in humans. Hepatology 32:11–16

Thomas E, Storb R, Clift RA et al 1975a Bone-marrow transplantation (first of two parts). N Engl J Med 292:832–843

Thomas ED, Storb R, Clift RA et al 1975b Bone-marrow transplantation (second of two parts). N Engl J Med 292:895–902

Vassilopoulos G, Wang PR, Russell DW 2003 Transplanted bone marrow regenerates liver by cell fusion. Nature 422:901–924

Wagers AJ, Sherwood RI, Christensen JL, Weissman IL 2002 Little evidence for developmental plasticity of adult hematopoietic stem cells. Science 297:2256–2259

Wang X, Montini E, Al-Dhalimy M et al 2002 Kinetics of liver repopulation after bone marrow transplantation. Am J Pathol 161:565–174

Wang X, Foster M, Al-Dhalimy M, Lagasse E, Finegold M, Grompe M 2003a The origin and liver repopulating capacity of murine oval cells. Proc Natl Acad Sci USA 100: 11881–11888

Wang X, Willenbring H, Akkari Y et al 2003b Cell fusion is the principal source of bone-marrow-derived hepatocytes. Nature 422:897–901

Willenbring H, Bailey AS, Foster M et al 2004 Myelomonocytic cells are sufficient for therapeutic cell fusion in liver. Nat Med 10:744–748

Yim AP 1982 Some flow-cytofluorimetric studies of the nuclear ploidy of mouse hepatocytes: III further observations on early changes in nuclear ploidy of mouse hepatocytes following various experimental procedures. Br J Exp Pathol 63:458–461

Ying QL, Nichols J, Evans EP, Smith AG 2002 Changing potency by spontaneous fusion. Nature 416:545–548

DISCUSSION

Gearhart: What percentage of liver function is required for survival — can we work out a target number of cells? And what would it be for the kidney?

Grompe: The lowest hanging fruit for the liver is haemophilia. This is why so many people have worked on gene therapy for haemophilia. It is estimated that 1% constitution with healthy blood clotting factor producing hepatocytes would be therapeutic. We also have data for PKU that the magic number is around 2%. In the urea cycle disorders we believe it to be about 50%. For the kidney, my guess would that be that it would have to be extensive to be therapeutic.

Weissman: I was looking at the kidney tubules and it looked like it was all or none.

Grompe: From endogenous tubule regeneration, we know one surviving cell within the tubule can replace everything in there. It is all or nothing.

Soria: My understanding is that there are around 30 different cell types in the kidney. You have shown tubule repopulation. What about all the other cell types?

Grompe: Our model is only set up to select for proximal tubule cells. We don't know whether other renal cell types are being replaced because we haven't looked at that. There is no reason to believe that there would be selection for donor cells of any other renal cell types.

Nadal-Ginard: In the liver did you only regenerate hepatocytes?

Grompe: Yes.

Lovell-Badge: You said you have complete reprogramming and you are selecting for function. Do you know from looking at other assays, such as RNA profiling, that you really are reprogramming everything?

Grompe: In a strict sense we have not done this. Obviously, we are using syngeneic hosts and donors. Other than for one gene, *Fah*, we have no formal proof that the hepatocellular genes are being expressed from the macrophage nucleus. We have just done experiments like that where we have used extensively genetically different transplants in immunodeficient animals to get at the question of whether there is really complete reprogramming. On the basis of the indirect evidence, I would be very surprised if it was less than complete.

Jaenisch: With all this fusion going on, which phenotype is dominant? Why don't you get macrophages? In fusions we know that embryonic stem (ES) cells are predominant over differentiated cells, but when you have two differentiated cells what determines which phenotype dominates? Is it a selection process?

Grompe: In the case of liver to macrophage, we are not sure who wins most of the time.

Jaenisch: Are there macrophage gene expression patterns in your fused liver cells?

Grompe: No, not after the completed selection. We have looked at multiple haematopoietic-specific genes after completed selection and we don't see any expression. However, immediately after fusion and before selection the situation may be different, and we are beginning to look at this.

Nadal-Ginard: What is the limit of expansion of these clonal fused cells?

Grompe: With hepatocytes we have published an experiment in which we did serial transplantation. This is why we called hepatocytes 'unipotential stem cells'. There is virtually no upper limit to their growth.

Nadal-Ginard: In a single mouse you should get progressive replacement by the one that has selective advantage. Like the bone marrow, you should end up with 90% of the hepatocytes being clonal.

Grompe: That is what occurs. There is one extreme example from retroviral marking where we have a whole liver reconstituted with one starting colony.

Weissman: I am going to propose an experiment. One of the questions that you would like to answer is whether this kind of cell fusion could be robust and whether it could be functional. That is, would it fit the idea that we normally make macrophages not only to clear up the mess, but also to rejuvenate tissues? It would be nice if you could do this experiment. You would like to have a chimeric liver where a significant number of *Fah*$^-$ cells are coexisting with *Fah*$^+$ cells from the beginning. If you put *Fah* double knockout ES cells into a *Fah*$^+$ blastocyst, then if you had sufficient markers you could score over time how much rejuvenation of the *Fah*$^-$ cells occurs. The only problem is the fusion that occurs with ES cells.

Grompe: I don't think that fusion is normally a physiological mechanism for tissue replacement. I believe this is a rare process that happens normally stochastically. What we are talking about is enhancing it artificially for therapeutic purposes.

Sullivan: I was interested in your assessment of cell fusion with the *LacZ*. What promoter was driving the *LacZ* transgene?

Grompe: It is Phil Soriano's Rosa26, which is ubiquitously expressed at a low level in every cell type.

Sullivan: Did you ever try taking cells from your donor and recipient and fusing them artificially to see whether there was extinction of *LacZ* expression? If this occurs then you wouldn't see it and you'd assume that cell fusion was lower than it might be.

Grompe: We haven't done that. We are just beginning the induced fusion experiments.

Sullivan: There is another interpretation of the data where your cell line showed that there was more fusion. In some of the *in vitro* work I have done it seems that when we stress cells they become more fusogenic. Perhaps the process of how we deal with the cells might make them more fusogenic.

Grompe: Currently this is a primary observation. Our interpretation has been that it is because more macrophages are going to the damaged tissue, but it is also possible that some kinds of stress induce high frequencies of cell fusion.

Weissman: What kind of anaesthetic did you use? A liver-toxic anaesthetic?

Grompe: We use a gas, which at least in the short term isn't hepatotoxic.

Gurdon: Is it possible that, for this rare event of reprogramming, you require a genetic change such that a very small fraction of the cells are actually different from the starting population. It could be only those cells that have acquired a capacity to undergo this change.

Grompe: It's possible, and it would be nice if this were the case: if one could identify and enrich those cells they would be a nice therapeutic vehicle. We are basically taking the approach that we would like to force cells to go that way by forced expression of fusogenic proteins, but there might be a subpopulation of naturally fusogenic macrophages.

Weissman: Certainly, in the bone marrow the osteoclasts are derived from cell fusion. Whether this is a particular differentiation stage of a pre-osteoclast or the normal function of the macrophage is unclear.

Grompe: There is genetic control of fusion. In myoblast fusion the satellite cell becomes a myotube and this is clearly not a stochastic process: it is a genetically and biochemically driven control mechanism. The syncytial trophoblast formation is caused by activation of the Herv-w endogenous retrovirus. There is clearly precedent for genetically induced cell fusion physiologically.

Gurdon: If you could find out what this fusogenic factor is, it would be of great interest and use.

Trounson: The reduction division resulted in a lot of cells that looked abnormal by chromosome analysis. Do the chromosomes get mixed? And are the tetraploid cells more stable and less aberrant when you examine them?

Grompe: We are just beginning to study this in more detail. One of the possibilities in our model is that the cytogenetic abnormalities aren't due to the fusion and reduction division, but the *Fah* mutant hepatocytes tend to be cytogenetically abnormal to begin with, so that the toxin that accumulates induces microtubule abnormalities and spindle failure. We are setting up experiments to see what happens in cell fusion without this toxin around. There are multiple translocations in this karyotype. It is hard to imagine that translocations would be due to a spindle abnormality or mis-seggregation. Clearly these cells have undergone some other kind of damage. I'm hopeful that they'll be cytogenetically normal in other systems.

Trounson: So you haven't determined whether their diploid daughter cells are a non-functional cast off population.

Grompe: I would say that the diploid–diploid fusions are more stable. We don't see any triploid hepatocytes. If the hexaploid hepatocytes had undergone a reduction division we would expect triploid hepatocytes, and we don't see these.

Goldman: With regards to fusion among different cell types, heterokaryon formation appears to have been reproducibly demonstrated in several cell types that may be preferentially susceptible to fusion. The repeated demonstrations of cell fusion with Purkinje cells, macrophages and hepatocytes suggests a considerable amount of cell type specificity in the ability of cells to fuse with a heterologous partner. If you were doing a Venn diagram of expressed transcripts in that setting, then a gene ontology search looking for surface antigens that are shared between the two might be expected to identify molecules that specifically permit fusion by these cells, as opposed to others less likely to fuse. Have any such molecules been identified yet, or have approaches like these been contemplated for picking out endogenous fusion partners at the gene or protein level?

Grompe: These are very early days. We have only known that these are macrophages for a couple of months. We suspected this, and this is why we did the experiments the way we did.

Goldman: Presumably there were some very specific surface recognition molecules that were subserving this process.

Grompe: This gets back to John Gurdon's question as to whether it is a subset of macrophages that are somehow ready to do this.

Goldman: I am thinking about the donor side, too. Within the nervous system the same molecules that mediate fusion by stromal cells and macrophages might be expressed by the Purkinje cells.

Weissman: There is one way to ask whether the fusogenic process of macrophages relates to their function of phagocytosis. I believe that the SIRP receptor on macrophages is signalled by CD47 not to be able to phagocytose the attached target. So you could conceivably do an experiment where you overexpress CD47 on liver cells and ask whether your population of macrophages or GM progenitors could still fuse in.

Grompe: We are actually more interested in strategies that would enhance the effect.

Nadal-Ginard: Is your fusion the same in the *Fah* mutant and in the wild-type? Why do you assume that the subpopulation that fuses is selected for by the macrophage rather than being a subgroup of the hepatocytes?

Grompe: The reason I think it is the same in *Fah* mutant as in other animal models is that other groups have looked at bone marrow-derived hepatocytes and the frequencies have been very similar. We are just selecting for these events in the *Fah* knockout and are not getting more fusion to begin with. We have showed that we cannot see fusion between hepatocytes themselves, which supports this. Is there a fusogenic subpopulation of hepatocytes? This is possible, and would be very interesting. We know from the cytogenetics that at least in terms of ploidy we get diploid–diploid and diploid–tetraploid fusions, so we don't see evidence that it is really limited in this particular approach.

Gearhart: Can you take a couple of minutes to talk about the progenitor-independent regeneration? What is known about this?

Grompe: This is from work by Snorri Thorgeirsson, who works on liver stem cells and liver cancer. He coined the term 'unipotential stem cell' for hepatocytes. These were experiments done in the 1970s in which people did serial partial hepatatectomy. One poor rat had a total of eight sequential partial hepatectomies. Each time there was complete hepatic reconstitution. We initially did serial transplantation in the *Fah* knockout with the idea that we would serially transplant only stem cells. Basically the differentiated cells unable to regenerate well would be diluted out by this process. We found that large, binucleated hepatocytes could be serially transplanted and we stopped after 100 cell doublings. This is roughly about the same number of cell doublings that has been shown for single haematopoietic stem cells. It appears as if this particular differentiated cell type can divide like an ES cell.

Solter: You also commented that depending on rat strain, 24–36 h after partial hepatectomy 100% of the cells are replicating DNA. Practically every single liver cell left is getting ready to divide.

Grompe: When I say that they divide like ES cells, interestingly telomerase does not come up. We are now designing experiments where we are trying to ask whether these unipotential stem cells are actually an artefact of the long telomeres of *Mus musculus*. It is not inconceivable that species with shorter telomeres such as humans are more dependent on progenitors than mice and rats.

Weissman: You can exhaust the telomeres in mouse haematopoietic stem cells. Even if you provide them with constitutively high level expression of telomeres they still have another mechanism of senescence that occurs after about 4 or 5 transplant generations.

Grompe: There are differentiated cell types that clearly don't need stem cells.

Nadal-Ginard: Are oval cells liver stem cells?

Grompe: No, they are committed progenitors that are bipotential and have some proliferation capacity. They are the daughters of the stem cells.

Nadal-Ginard: Which are the real stem cells?

Grompe: That's what we would like to know. Anatomically, they are probably in the Canal of Hering.

Gearhart: Returning to a topic we discussed earlier, your definition of successful ES cell therapy is that the cell has grafted in and directly substituted or replaced a cell that has been lost or damaged; that the grafted cell has done something directly.

Grompe: Yes. My understanding is that the ES cell derived HSCs that George Daley has been working with have some properties of HSCs but they are not the real thing.

Weissman: I think they are equivalent to the yolk sac blood islet stage of haematopoiesis. If you transplant yolk sac blood islets into adults they don't take.

But if you transplant them into the same age yolk sacs (day 8) they always contribute to lifelong haematopoiesis. The cells they give rise to in adults look just like robust bone marrow-derived HSCs. Obviously, they are going through changes as they move from the yolk sac injection site to the body. We know already that part of this movement is dependent on $\alpha4\beta1$ integrin partners, because these knockouts prevent the development of adult haematopoiesis, and $\alpha4\beta1$ is the homing receptor that transfused HSCs from the marrow use to get back to the marrow. If there are analogous issues with other tissue developments, then you might expect to encounter problems. That is why we shouldn't get focused on therapeutic cloning as the sole reason for this research, and we should be careful not to give the impression that we will have therapies in the near term. If we start out, I think eventually we will, because there is a reductionist path in front of us.

Trounson: I want to ask Markus about the muscle satellite cells and the fact that they are supposed to arise from some kind of bone marrow-derived cell. What is your view about this population?

Grompe: Not being a muscle biologist, this is confusing. We are not doing this work in my own lab. There is strong evidence that there are satellite cell precursors in the bone marrow. My reading of the data is that it is not the haematopoietic stem cell. This is the way I would summarize my understanding of it.

Weissman: Part of the issue is, are all satellite cells truly muscle stem cells, or are they a heterogeneous population? When we tried to repeat some studies that said that satellite cells could be derived from haematopoietic stem cells, we couldn't replicate them. But as we investigated further, the criteria that they are visually there and have particular gene expression, and are robust precursors of muscle cells kept arising. This needs a little more clarity.

Gearhart: One of the dangling questions from the previous discussion was whether the multipotent adult progenitor cells (MAPCs) form tumours. Catherine Verfaillie, can you comment?

Verfaillie: So far we have not seen any. I can't say they don't though. We have given them i.v. and subcutaneously in large doses. We have given them mainly to non-SCID mice. It is possible that they might form tumours so we are now doing studies in nude mice.

Minger: I'm not sure that the forming tumours that contain all germ lineages is an absolute criterion for demonstrating pluripotency. Nico Heins' group in Sweden has generated six ES cell lines that are ES cells by all criteria (Heins et al 2004). When they have transplanted them into SCID mice only four of them make solid tumours and two of them make cystic tumours that express only two of the germ layers. What does this say about pluripotency?

Solter: With mouse ES and EC cells some cell lines become restricted in terms of what kinds of tumours they make.

Trounson: That happens with EC cells. There is an adaptation to culture such that you can't readily direct differentiation into some lineages. You have to be careful about passage time when you are discussing these cells.

Gearhart: With the EG cells that we have we have not been able to show teratoma formation. Yet we can derive a large variety of cell types from them. We don't know the significance of this.

Reference

Heins N, Englund MCO, Sjöblom C et al 2004 Derivation, characterization, and differentiation of human embryonic stem cells. Stem Cells 22:367–376

Normal and neoplastic stem cells

Irving L. Weissman

Department of Pathology, Stanford University Medical Center, B257 Beckman Center, Stanford, CA 94305, USA

Abstract. Stem cells are cells that at the single cell level both self-renew and give rise to differentiated progeny. Self renewal is the property that distinguishes stem cells and progenitors, and in the blood-forming system explains why haematopoietic stem cells (HSCs), not progenitors, are the only cells capable of providing rapid and sustained regeneration of the blood-forming system after ablation by cancer chemo- and radiotherapies. Cancer-free prospectively purified HSCs regenerate the haematopoietic system of patients as rapidly as a marrow or mobilized blood transplant, but without the risk of re-seeding the body with cancer cells. Further, purified allogeneic HSCs can establish donor-specific tolerance to subsequent tissue grafts. However, in contrast to widely-publicized reports of HSC plasticity, we have not been able to show transdifferentiation of HSC to muscle, heart, brain or gut, and conclude that rare cell fusions and incomplete purifications are likely explanations for the other published results. The ability to self-renew is also potentially dangerous, as poorly regulated self renewal is, we believe, a central lesion in all cancers. We have recently shown that myeloid leukaemias in mouse and human are often driven by rare leukaemia (cancer) stem cells which are at the progenitor stage of differentiation, but have activated the self-renewing cell division pathway normally used only by HSCs. Similar cancer stem cells have been isolated in other tumours.

2005 Stem cells: nuclear reprogramming and therapeutic applications. Wiley, Chichester (Novartis Foundation Symposium 265) p 35–54

The only treatment for haematopoietic failure following whole body radiation is transplantation of bone marrow cells, or in fact, the haematopoietic stem cells (HSCs) that are responsible entirely for rapid and sustained regeneration of the blood-forming system in these hosts (Gengozian et al 1957, Ford et al 1956, Nowell et al 1956, Weissman 2000a, Blume et al 2004). The haematopoietic system is destroyed by the minimal lethal doses of X- or nuclear radiation, and also by chemotherapeutic agents which, like radiation, largely kill dividing cells. In the 1960s Wu, Becker, Till, Siminovitch and McCulloch showed that rare cells in bone marrow were clonogenic haematopoietic cells that gave rise to all myeloerythroid cells, and sometimes self-renewed (Till & McCulloch 1961, Becker et al 1963, Siminovitch et al 1963, Wu et al 1963), and also gave rise to lymphoid cells. These genetic marking experiments established the fact that there must exist in bone marrow single cells that can both self-renew and generate all of the populations of cells in the blood; these cells were called at that time pluripotent HSCs, a term which has later been modified to multipotent HSCs (Weissman 2000b, Spangrude et al 1988). But knowing that stem cells exist by retrospective analysis of randomly gene marked cells is not the same as having pure populations of HSCs for study that can be prospectively isolated.

To search for those cells in bone marrow that contain the activity of HSCs we developed assays for the clonal precursors of T lymphocytes (Ezine et al 1984) and of B lymphocytes (Whitlock et al 1987, Muller-Sieburg 1986) in addition to the clonal precursors of myeloerythroid cells found in spleen colonies (Till & McCulloch 1961, Becker et al 1963, Siminovitch et al 1963, Wu et al 1963). At the point we began, it was already known that the spleen colony-forming cells that Till and McCulloch studied were in fact probably the outcome of oligopotent myeloerythroid progenitors rather than stem cells, and that they were different from colonies that arise at about 14 days—close to the time of death—which come from more primitive haematopoietic cells (Magli et al 1982). For each of those assays a single colony could be derived from 10^3 to 10^4 cells, so we had a quantitative assessment for the enrichment of these clonal precursors of T, B and myeloerythroid cells. We then needed a way to fractionate prospectively cells from the bone marrow to put into these colony assays. We chose to produce large numbers of monoclonal antibodies (Kohler & Millstein 1975) that detected subsets of cells found in bone marrow, and to use high speed fluorescence-activated cells sorters (FACS) to purify these cells (Hulett et al 1969). In 1986 we reported a high degree of enrichment of multipotent HSCs, and in 1988 their full isolation (Spangrude et al 1988, Muller-Sieburg 1986). At that point mouse haematopoietic multipotent marrow cells represented 1 in 2000 cells in the young adult mouse marrow bones, and were 2000-fold enriched for the ability to radioprotect lethally irradiated hosts by donor-derived reconstitution of all blood cell types for life (Spangrude et al 1988). There were no other cells than these cells in bone marrow capable of long term multi-lineage reconstitution (Uchida & Weissman 1992), and at the single cell levels these were all multipotents, although some reconstituted for weeks, or a few months, and others reconstituted for life (Smith et al 1991, Wagers et al 2002, Uchida 1992). We later showed that these mouse multipotent cells in fact could be subdivided into three subsets—long term HSCs (LT-HSCs), short term HSC (ST-HSCs) and multipotent progenitors (MPPs), and that these cells existed in a lineage (Morrison & Weissman 1994, Morrison et al 1997a, Christensen et al 2001). Of these only LT-HSCs self-renewed in apparent perpetuity, while ST-HSCs had a programmed self-renewal lifespan of 6–8 weeks, and multipotent progenitors, of less than two weeks (Morrison & Weissman 1994, Morrison et al 1997a, Christensen & Weissman 2001). The search for human HSCs didn't take long.

Using similar assays for clonogenic, lymphoid and myeloerythroid progenitors it was possible to isolate candidate human haematopoietic and multipotent stem and progenitor cells, with their own distinctive and similar cell surface markers (Baum et al 1992, Uchida et al 1998a, Bhatia et al 1997). We demonstrated long-term multilineage reconstitutive activity with these, but no other cells in human haematopoietic tissues by using them to reconstitute human fetal bone, human fetal thymus, human fetal liver and human fetal lymphoid organs implanted into fully immunodeficient SCID-hu mice (Baum et al 1992, McCune et al 1988, Tsukamoto et al 1995). Later, it was shown that these cells too could be divided into LT-HSCs and shorter-lived multipotent stem and progenitors (Uchida et al 1998a).

Clinical trials using purified human HSCs

Modern cancer chemotherapy usually involves simultaneous administration of three or four independent chemotherapeutic agents, plus or minus local or whole body radiation. For many cancers, however, the dose to kill all existing cancer cells in the body is at a level which kills haematopoietic stem cells down to the level that regeneration is not possible before death. Autologous haematopoietic cell transplant patients were treated by removing their bone marrow before these myeloablative therapies are instituted, freezing it, and rescuing them after the chemotherapy administration is finished (for review see Thomas 1999). However, for many late stage cancers and leukaemias the spread of metastatic cancer cells also contaminates blood and bone marrow. It makes little sense to go to the trouble to destroy all cancer cells within the body, if you follow that with re-seeding of the body with marrow transplants or mobilized blood transplants containing cancer cells. We were able to demonstrate that FACS sorting of human haematopoietic stem cells using marker antibodies resulted in at least a 100 000-fold depletion of contaminating cancer cells (breast cancer cells, or non-Hodgkin's lymphoma cells, or multiple myeloma cells) (Manz et al 2004). Thus we could for the first time deliver back to the patient HSCs with few or no contaminating cancer cells from their own bodies. Three phase 1/2 clinical trials were carried out with prospectively isolated human HSCs: stage IV (widely metastatic) breast cancer (Negrin et al 2000), non-Hodgkin's lymphoma (Vose et al 2001), and multiple myeloma (Michallet et al 2000). In all three trials the purified HSCs were given at doses that enabled rapid return of white cells and platelets, comparable to the time of regeneration of these blood elements by mobilized blood (Negrin et al 2000, Vose et al 2001, Michallet et al 2000). At about four years post transplantation about 35–40% of Stage IV breast cancer treatment patients receiving this treatment were alive without evidence of disease, and about 55% of the non-Hodgkin's lymphoma were alive without evidence of disease (R. Negrin, personal communication; Michallet et al 2000). In different trials using unpurified mobilized peripheral blood, the four-year Stage IV breast cancer survival without disease progression was about 6% (Stadtmauer et al 2000). Because these were not experiments where patients were prospectively randomized into pure stem cell versus unmanipulated mobilized blood transplants, it is not possible to tell whether these increased cancer-free survival rates are significant and reproducible, so further trials are warranted.

Transplantation of allogeneic HSCs leads to donor-specific tolerance of tissue grafts, and donor-specific abrogation of autoimmune diabetes in mice

Transplantation of allogeneic marrow or mobilized blood between individuals, even if they are matched at the major histocompatibility locus (Weissman 1963, Main & Prehn 1955), leads to a serious complication: T cells from the donor cause graft versus host disease (GVHD) (for review see Ferrara et al 1999). Purification of HSCs in mouse and human eliminates contaminating T cells. We have carried out allogeneic HSC transplants in different mouse strain combinations. Higher doses of HSC were required in the allogeneic mismatched transplants than in syngeneic transplants to achieve rapid and sustained engraftment (Shizuru et al 1996, 2000, Uchida et al 1998b). Such chimeras were permanently and specifically tolerant of donor tissue and organ transplants, whether the organ transplants were given the same day as HSCs, or up to a year later (Shizuru et al 2000, Gandy & Weissman 1998). This method of inducing transplant tolerance should eliminate both GVHD and chronic host immunosuppression.

These mouse preclinical experiments also allow one to assess whether various genetically defective haematolymphoid systems and hosts can be replaced by healthy blood-forming systems. These include not only genetic defects such as aplastic anaemia, thalassaemia, severe combined immunodeficiency and sickle cell anaemia, but also replacement of immune systems prone to attack self that create autoimmune diseases (Weissman 2000a). These autoimmune diseases include Type 1 (juvenile) diabetes (autoimmune T cells destroy insulin producing pancreatic islets) (Gershenfeld & Weissman 1986, Wood et al 1988, Held et al 1990), multiple sclerosis (autoimmune T cells destroy myelinating oligodendrocytes), rheumatoid arthritis (Griffiths et al 1992), systemic lupus erythematosus, and ankylosing spondylitis, to name a few of the genetically determined autoimmune disorders. We have shown that normal HSCs can replace the autoimmune- prone haematolymphoid system of mice (NOD) with developing Type 1 diabetes at the stages in which polydypsia and polyuria precede complete islet destruction; these mice are cured of the development of this disease (Beilhack et al 2003). Mice already having complete destruction of the islets can also have the autoimmune part of their disease abrogated with HSC transplants from donors lacking the genetic risk for diabetes, and in some cases frank diabetic animals treated with HSCs and islet transplants from healthy donors are also cured of their need for insulin and the complications of their disease (Beilhack et al 2003).

Expansion of HSCs

The number of HSCs that one can isolate from mobilized blood, or from umbilical cord, or from bone marrow limits the full application of HSC transplantation in human, whether in response to accidental or intentional nuclear radiation exposure, or transplantation in the treatment of diseases as described above. Attempts to expand HSCs with the known cytokines, stem cell factor/steel factor (SLF), thrombopoietin (TPO), interleukins 1, 3, 6, 11 plus or minus the myeloerythroid cytokines GMCSF, GCSF, MCSF and erythropoietin have never resulted in a significant expansion of HSCs. Rather, they induce many, if not most HSCs into cell divisions which are accompanied by cell differentiations (Domen & Weissman 2000). Yet there are many experiments wherein single or a few HSCs were transplanted into animals, and in those settings animals expanded the number of HSCs at least 10 000-fold at the steady state while permitting the daughters of HSCs to regenerate full blood-forming systems (Smith et al 1991, Wagers et al 2002, Uchida 1992, Morrison & Weissman 1994). We have recently found a pathway that enables mouse HSCs to undergo self-renewing cell divisions, with progeny that are functional in haematopoietic cell transplantations into lethally irradiated hosts (Reya et al 2003, Willert et al 2003). LT-HSCs express elements of the Wnt/fzd/β-catenin signalling pathway (E. Ranheim, S. Prohaska, C. Forsberg, A. Wagers, K. Li, S. Cheshier and I. Weissman, unpublished results). The interaction between Wnt, Fzd and the coreceptors LRP6 resulted in the release in the cytoplasm of activated, unphosphorylated β-catenin, which upon translocation to the nucleus, bound to the DNA binding proteins LEF/TCF, converting them from repressors of transcription to activators of transcription (Nelson & Nusse 2004). Proteins in the multiprotein destruction complex are negative regulators of β-catenin activation and stimulation of LEF/TCF transcription, and it has been found empirically that destructive mutations of each of those negative regulatory elements can play a role in the development of at least colorectal carcinomas (Fearon & Vogelstein 1990). We then tested the possibility that the Wnt/Fzd/β-catenin pathway is a regulator of self-renewing divisions of at least mouse HSC, and demonstrated that addition of highly purified Wnt3A to HSC leads to their expansion, as does transfection of activated β-catenin genes into HSCs (Reya et al 2003, Willert et al 2003). HSC cell lines requiring no serum could be established that expanded 100- to 1000-fold, and these expanded HSCs were also transplantable (Reya et al 2003); such tissue culture expansions of HSCs could only be accomplished if these cells were blocked in the programmed cell death pathway by BCL-2 (Domen et al 1998, 2000). Wnt activation of HSCs leads to up-regulation of other genes implicated in HSC self renewal, including Notch 1 and HOX B4 (Reya et al 2003, Antonchuk et al 2002, Varnum-Finney et al 2000).

HSCs normally traffic from bone marrow to blood

The optimized mobilization protocol of cytoxan plus G-CSF results in every resident LT-HSC in mouse bone marrow undergoing several self-renewing cell divisions, expanding the number of HSCs 12- to 15-fold in a matter of two to three days (Morrison et al 1997b, Wright et al 2001a). It appears that on the second or third day following mobilization, up to one-half of the daughter cells of self-renewing dividing LT-HSCs leave the bone marrow, enter the blood, and within minutes engraft other bone marrow, spleen or even liver haematopoietic sites (Wright et al 2001a). In normal mice, transfused LT-HSCs rapidly emigrate from blood to haematopoietic tissues (Morrison et al 1997b). To maintain the steady state of 100 LT-HSCs the bone marrow must produce and release between 10^4 and 10^5 LT-HSCs per mouse per day into the blood stream (Wright et al 2001b). Resident HSC cells in the blood enter empty haematopoietic niches elsewhere in the bone marrow, and provide sustained haematopoietic stem cell self-renewal and haematopoiesis (Wright et al 2001b). Concurrent with the discovery of this massive HSC flux through all tissues were the early *claims* that brain cells in developed animals contained stem cells that could turn into blood-forming stem cells (Bjornson et al 1999), that bone marrow blood-forming cells could give rise to neurons in the brain (Brazelton et al 2000, Mezey et al 2000), skeletal myocytes in muscle (Morrison & Weissman 2004, Morshead et al 2002, Muller-Sieburg et al 1986), regenerating myocardium and blood vessels in the heart (Sata et al 2002, Orlic et al 2001a). These latter studies were not carried out in a way that would distinguish between tissue specific stem cells coexisting with itinerant HSCs in the tissues. These experiments were the basis of claims that demonstrated the *plasticity* of adult stem cells.

Can blood-forming stem cells or other resident bone marrow cells transdifferentiate to other tissue specific stem and progenitor cells, or can they contribute to regeneration of non-haematopoietic damaged tissues?

We had called into question the claims that one adult tissue-type stem cell could turn into another tissue-type cell (Anderson et al 2001, Weissman et al 2001). We proposed that one could only claim one tissue could turn into another if one began with highly purified, well known tissue-committed stem cells, ideally single stem cells (for example, HSCs), transferred to genetically distinct hosts, and demonstrated the progeny of that stem cell included both its normal tissue derivatives as well as other kinds of tissues (blood and other tissues in the HSC case). The existence of large numbers of circulating haematopoietic stem cells through each tissue requires this kind of single cell analysis, or purification to homogeneity of cells. The initial claim that skeletal muscle contained a common stem cell for blood and muscle (Jackson et al 1999), was later amended when this second criterion was applied, and it turned out that the cells giving rise to the blood and blood forming system were HSCs in muscle with HSC markers, whereas the population giving rise to muscle were cells within muscle lacking haematopoietic lineage markers, and that could give rise to muscle only (McKinney-Freeman et al 2002). There was a second claim that neurospheres derived from a clonogenic brain cell could give rise *in vitro* to brain cells, but upon injection *in vivo*, and after a delay of about seven months, could replace the full haematopoietic system of sublethally irradiated allogeneic hosts (Bjornson et al 1999). This failed replication (Morshead et al 2002).

In another experiment single visually observed lateral ventrical ependymal cells were isolated, cultured extensively to produce neurospheres, and the daughter cells placed in the blastocyst of the same species (mouse) or directly injected into a developing chick embryo; donor markers could be found in a wide variety of tissues, though oddly excluding haematopoietic tissue (Clarke et al 2000), different than tissue produced from another neurosphere experiment (Bjornson et al 1999). There are several experiments wherein bone marrow cells, often characterized as haematopoietic stem cells but without the attendant purification, injected into lethally irradiated animals give rise to donor derived myofibres in muscle and cortical neurons as well as cerebellar Purkinje cells in the brain (Brazelton et al 2000, Mezey et al 2000, Ferrari et al 1998, Gussoni et al 1999, Jackson et al 2001, Orlic et al 2001a, Sata et al 2002). We cannot reproduce the demonstration of production of cortical neurons from marrow or haematopoietic stem cells precursors (M. Massengale, A. Wagers and I. L. Weissman, unpublished data). However, donor markers can appear in muscle, in Purkinje cells in the cerebellum, rarely in myocardial epithelium, and in liver cells (Orlic et al 2001a,b, Lagasse et al 2000, Wang et al 2003 and see below). There are experiments wherein bone marrow cells and cells enriched but not highly purified for haematopoietic stem cells injected into the injured myocardium shortly after a left anterior descending coronary ligation has been carried out, and the donor cells are claimed not only to give rise to cardiac muscle, blood vessel, and perivascular smooth muscle cells containing donor markers, as well as providing a functional regeneration of the heart (Orlic et al 2001a,b). These experiments have led to extensive clinical trials in humans (Caplice & Gersch 2003, Strauer et al 2002, Stamm et al 2003, Belenkov et al 2003, Bel et al 2003). We have carried out extensive experiments attempting to repeat these findings, and always find that either haematopoietic stem cells or bone marrow cells having donor markers can be found in the inflamed and damaged cardiac muscle, but none of those cells express the markers, the morphology, nor the function of cardiac muscle, or of cardiac blood vessels, or of cardiac smooth muscle cells (Balsam et al 2004).

Cell fusion and claims of stem cell plasticity

The fusion of two cells to produce a heterokaryon, a cell with two different nuclei, is a rare event in nature and in pathology, but it has been recorded at several different levels. First, in many pathological states following infection with agents as widespread as Sendai virus or *Mycobacterium tuberculosis*, multinucleated giant cells derived from the fusion of cells of the monocyte/macrophages series are a constant and often pathognomonic sign of the infection type (Langhans 1868). Second, cells of the monocyte/macrophage series normally fuse by the thousands every day in every bone to form osteoclasts, cells whose phagocytosis and digestion processes remodel the bones to make space for bone marrow (Vignery 2000). Third, the Sendai virus from Sendai virus pneumonia has been used experimentally to produce artificial heterokaryons since the mid 1960s (Harris 1964, 1965, Harris et al 1965). In the normal development of all skeletal muscles, myoblast progenitors fuse to each other or to existing myotubes to form multinucleated skeletal muscle cells (Abmayr et al 2003), a physiological phenomenon. True embryonic stem cells line derived from mouse inner cell mouse blastocysts have been shown when mixed together *in vitro* to fuse as well (Ying et al 2002, Terada et al 2002). It is unclear which of these cell fusions is physiological and which pathological, especially when cells of the monocyte macrophages is involved, whose normal function is to adhere to dying and dangerous cells during the process of phagocytosis. There are several well documented cases wherein stem cells or tissues containing stem cells can, upon *in vivo* injection, give rise to the donor markers in cells of other tissue types. For example, purified HSCs could contribute donor markers as well as a functional enzyme to highly damaged liver cells in a host lacking FAH (Lagasse et al 2000). A systematic analysis of the chromosomes in these regenerated livers have demonstrated that in every case there was a fusion between host and donor haematopoietic cell progeny (Wang et al 2003, Alvarez-Dolado et al 2003). Mice restored with highly purified HSCs, some with a single HSC that gives rise to haematopoiesis, or reconstitute with whole bone marrow containing that number of stem cells used in the HSC reconstitution were tested with and without cardiotoxin and crush damage to skeletal muscles in a variety of muscles to test for donor markers in skeletal muscle (Sherwood et al 2004). Rare donor marked myocytes were found and all were donor/host cell fusions. Thus rare cell fusions in for the most part highly damaged muscles do come from blood-borne circulating precursors (Sherwood et al 2004).

Do other tissues have tissue specific stem cells used in their generation and regeneration throughout life?

In addition to HSCs, the following stem cells have been prospectively isolated to homogeneity or isolated by cell culture: peripheral nervous system stem cells (Stemple & Anderson 1992, Morrison et al 1999) and CNS stem cells (Reynolds & Weiss 1992, Palmer et al 1997, Uchida et al 2000). In addition highly enriched populations that contain stem cells have been found for the skin (Watt 1998), as well as mesenchymal stem cells (Jiang et al 2002, Colter et al 2001, Short et al 2003, Caplan & Bruder 2001, Pittenger et al 1999). In the case of human CNS stem cells, extensive experiments using their transplantation into the lateral ventricles or into the brain or into the spinal cord of SCID mice have shown that they contributed in a robust way to engraftment of the neurogenerative cells (the subventricular zone of the lateral ventricles and the subgranular zone of the dentate gyrus in the hippocampus) in adult animals, as well as daughter cells that appear to be undergoing site-specific self-renewal, migration, and differentiation into oligodendrocytes, various neurons, and astrocytes in an apparent site-appropriate manner (Uchida et al 2000, Tamaki et al 2002).

We do not yet know that all organ systems are based on the stem→ progenitor →progeny model, but the method that led to the isolation of HSCs, peripheral nervous system stem cells, and central nervous stem cells is general enough to look for such tissue- and organ-specific stem cells. It should be added that the same method was used for prospective isolation of most if not all haematopoietic progenitor cells downstream of HSCs in mouse and human, and these cells too are capable of more limited and more specific regeneration of various important blood cell types, although none of these cells self-renew, and therefore all such transplants are transient in nature (Kondo et al 1997, 2003, Akashi et al 2000, Traver et al 2000, 2001, Manz et al 2001, 2002, Mebius et al 2001, Na Nakorn et al 2002, 2003, Miyamoto et al 2002, Arber et al 2003, BitMansour et al 2002).

Cancer stem cells

Cancerous tissues, which are obviously derived from normal tissues following somatic genetic and epigenetic changes, contain within them cancer or leukaemia stem cells, and for the most part these cancer and leukaemia stem cells differ from normal tissue stem cells (Sell & Pierce 1994, Rohrer et al 1977, Miyamoto et al

2000, Gale et al 1997, Ford et al 1998, Bonnet & Dick 1997, Reya et al 2001, Al-Hajj et al 2003, Lapidot et al 1994, Jamieson et al 2004, Cozzio et al 2003) as well as the non-malignant progeny they produce in a failed attempt to make normal tissue. These can include progenitor cells within tissues that have acquired the capacity for self renewal, as well as multiple genetic and epigenetic events to avoid programmed cell death, immune-mediated death and phagocytosis, and limitation of replication lifespans by loss of telomere protective sequences (Lagasse & Weissman 1994, Traver et al 1998, Brown et al 1997, Kogan et al 2001, Yuan et al 2001, Allsopp et al 2001, Morrison et al 1996, Park et al 2003). The subject of cancer and leukaemia stem cells is better described elsewhere (Reya et al 2001).

References

Abmayr SM, Balagopalan L, Galletta BJ, Hong SJ 2003 Cell and molecular biology of myoblast fusion. Int Rev Cytol 225:33–89

Akashi K, Traver D, Miyamoto T, Weissman IL 2000 A clonogenic common myeloid progenitor that gives rise to all myeloid lineages. Nature 404:193–197

Al-Hajj M, Wicha MS, Benito-Hernandez A, Morrison SJ, Clarke MF 2003 Prospective identification of tumorigenic breast cancer cells. Proc Natl Acad Sci USA 100:3983–3988 (Erratum in Proc Natl Acad Sci USA 100:6890)

Allsopp RC, Cheshier S, Weissman IL 2001 Telomere shortening accompanies increased cell cycle activity during serial transplantation of hematopoietic stem cells. J Exp Med 193: 917–924

Alvarez-Dolado M, Pardal R, Garcia-Verdugo JM et al 2003 Fusion of bone-marrow-derived cells with Purkinje neurons, cardiomyocytes and hepatocytes. Nature 425:968–973

Anderson DJ, Gage FH, Weissman IL 2001 Can stem cells cross lineage boundaries? Nat Med 7:393–395

Antonchuk J, Sauvageau G, Humphries RK 2002 HOXB4-induced expansion of adult hematopoietic stem cells ex vivo. Cell 109:39–45

Arber C, BitMansour A, Sparer TE et al 2003 Common lymphoid progenitors rapidly engraft and protect against lethal murine cytomegalovirus infection after haematopoietic stem cell transplantation. Blood 102:421–428

Balsam LB, Wagers AJ, Christensen JL, Kofidis T, Weissman IL, Robbins RC 2004 Haematopoietic stem cells adopt mature hematopoietic fates in ischemic myocardium. Nature 428:668–673

Baum CM, Weissman IL, Tsukamoto AS, Buckle AM, Peault B 1992 Isolation of a candidate human hematopoietic stem-cell population. Proc Natl Acad Sci USA 89:2804–2808

Becker A, McCulloch E, Till J 1963 Cytological demonstration of the clonal nature of spleen colonies derived from transplanted mouse marrow cells. Nature 197:452–454

Beilhack GF, Scheffold YC, Weissman IL et al 2003 Purified allogeneic hematopoietic stem cell transplantation blocks diabetes pathogenesis in NOD mice. Diabetes 52:59–68

Bel A, Messas E, Onnik A et al 2003 Transplantation of autologous fresh bone marrow into infarcted myocardium: a word of caution. Circulation 108:247II–252

Belenkov Iu N, Ageev FT, Mareev ViuSavchenko VG 2003 Mobilization of bone marrow stem cells in the management of patients with heart failure. Protocol and first results of ROT FRONT trial. Kardiologiia 43:7–12

Bhatia M, Wang JC, Kapp U, Bonnet D, Dick JE 1997 Purification of primitive human hematopoietic cells capable of repopulating immune-deficient mice. Proc Natl Acad Sci 94:5320–5325

BitMansour A, Burns SM, Traver D et al 2002 Myeloid progenitors protect against invasive aspergillosis and Pseudomonas aeruginosa infection following hematopoietic stem cell transplantation. Blood 100:4660–4667

Bjornson CR, Rietze RL, Reynolds BA, Magli MC, Vescovi AL 1999 Turning brain into blood: a hematopoietic fate adopted by adult neural stem cells in vivo. Science. 283:534–537

Blume KG, Forman SJ, Appelbaum FR (eds) 2004 Thomas' hematopoietic cell transplantation. 3rd edn. Blackwell Publishing, Malden MA

Bonnet D, Dick JE 1997 Human acute myeloid leukemia is organized as a hierarchy that originates from a primitive hematopoietic cell. Nat Med 3:730–737

Brazelton TR, Rossi FM, Keshet GI, Blau HM 2000 From marrow to brain: expression of neuronal phenotypes in adult mice. Science 290:1775–1779

Brown D, Kogan S, Lagasse E et al 1997 A PMLRARalpha transgene initiates murine acute promyelocytic leukemia. Proc Natl Acad Sci USA 94:2551–2556

Caplice NM, Gersh BJ 2003 Stem cells to repair the heart: a clinical perspective. Circ Res 92:6–8

Caplan AI, Bruder SP 2001 Mesenchymal stem cells: building blocks for molecular medicine in the 21st century. Trends Mol Med 7:259–264

Christensen JL, Weissman IL 2001 Flk-2 is a marker in hematopoietic stem cell differentiation: a simple method to isolate long-term stem cells. Proc Natl Acad Sci USA 98:14541–14546

Clarke DL, Johansson CB, Wilbertz J et al 2000 Generalized potential of adult neural stem cells. Science 288:1660–1663

Colter DC, Sekiya I, Prockop DJ 2001 Identification of a subpopulation of rapidly self-renewing and multipotential adult stem cells in colonies of human marrow stromal cells. Proc Natl Acad Sci USA 98:7841–7845

Cozzio A, Passegue E, Ayton PM, Karsunky H, Cleary ML, Weissman IL 2003 Similar MLL-associated leukemias arising from self-renewing stem cells and short-lived myeloid progenitors. Genes Dev 17:3029–3035

Domen J, Weissman IL 2000 Hematopoietic stem cells need two signals to prevent apoptosis; BCL–2 can provide one of these, Kitl/c–Kit signaling the other. J Exp Med 192:1707–1718

Domen J, Gandy KL, Weissman IL 1998 Systemic overexpression of BCL–2 in the hematopoietic system protects transgenic mice from the consequences of lethal irradiation. Blood 91:2272–2282

Domen J, Cheshier S, Weissman IL 2000 The role of apoptosis in the regulation of hematopoietic stem cells: overexpression of BCL–2 increases both their number and repopulation potential. J Exp Med 191:253–263

Ezine S, Weissman IL, Rouse RV 1984 Bone marrow cells give rise to distinct cell clones within the thymus. Nature 309:629–631

Fearon ER, Vogelstein B 1990 A genetic model for colorectal tumorigenesis. Cell 61:759–767.

Ferrara JL, Levy R, Chao NJ 1999 Pathophysiologic mechanisms of acute graft-vs.-host disease. Biol Blood Marrow Transplant 5:347–356

Ferrari G, Cusella De Angelis G, Colleta M et al 1998 Muscle regeneration by bone marrow-derived myogenic progenitors. Science 279:1528–1530 (Erratum in Science 281:923)

Ford AM, Bennett CA, Price CM, Bruin MC, Van Wering ER, Greaves M 1998 Fetal origins of the TEL-AML1 fusion gene in identical twins with leukemia. Proc Natl Acad Sci USA 95:4584–4588

Ford CE, Hamerton JLl, Barnes DW, Loutit JF 1956 Cytological identification of radiation chimaeras. Nature 177:239–247

Gale KB, Ford AM, Repp R et al 1997 Backtracking leukemia to birth: identification of clonotypic gene fusion sequences in neonatal blood spots. Proc Natl Acad Sci USA 94: 13950–13954

Gandy KL, Weissman IL 1998 Tolerance of allogeneic heart grafts in mice simultaneously reconstituted with purified allogeneic hematopoietic stem cells. Transplantation 65: 295–304

Gengozian N, Urso IS, Congdon CC et al 1957 Thymus specificity in lethally irradiated mice treated with rat bone marrow. Proc Soc Exp Biol 96:714

Gershenfeld H, Weissman IL 1986 Cloning of a cDNA T cell-specific serine protease from a cytotoxic T lymphocyte. Science 232:854–858

Griffiths G, Alpert S, Lambert E, McGuire J, Weissman IL 1992 Perforin and granzyme A expression identifying cytolytic lymphocytes in rheumatoid arthritis. Proc Natl Acad Sci USA 89:549–553

Gussoni E, Soneoka Y, Strickland CD et al 1999 Dystrophin expression in the mdx mouse restored by stem cell transplantation. Nature 401:390–394

Harris H 1964 Transfer of radioactivity from nuclear to cytomplasmic ribonucleic acid. Nature 202:249–252

Harris H 1965 Behaviour of differentiated nuclei in heterokaryons of animal cells from different species. Nature 206:583–588

Harris H, Watkins JF, Campbell GL, Evans EP, Ford CE 1965 Mitosis in hybrid cells derived from mouse and man. Nature 207:606–608

Held W, MacDonald HR, Weissman IL, Hess MW, Mueller C 1990 Genes encoding tumor necrosis factor and granzyme A are expressed during development of autoimmune diabetes. Proc Natl Acad Sci USA 87:2239–2243

Hulett HR, Bonner WA, Barrett J, Herzenberg LA 1969 Cell sorting: automated separation of mammalian cells as a function of intracellular fluorescence. Science 166:747–749

Jackson KA, Mi T, Goodell MA 1999 Hematopoietic potential of stem cells isolated from murine skeletal muscle. Proc Natl Acad Sci USA 96:14482–14486

Jackson KA, Majka SM, Wang H et al 2001 Regeneration of ischemic cardiac muscle and vascular endothelium by adult stem cells. J Clin Invest 107:1395–1402

Jamieson CHM Ailles LE, Dylla SJ et al 2004 Granulocyte-macrophage progenitors as candidate leukemic stem cells in blast-crisis CML. N Eng J Med 351:657–667

Jiang Y, Yahagirdar BN, Reinhardt RL et al 2002 Pluripotency of mesenchymal stem cells derived from adult marrow. Nature 418:41–49

Kogan SC, Brown De, Schultz DB et al 2001 BCL-2 cooperates with promyelocytic leukemia retinoic acid receptor alpha chimeric protein (PMLRARalpha) to block neutrophil differentiation and initiate acute leukemia. J Exp Med 193:531–543

Kohler G, Milstein C 1975 Continuous cultures of fused cells secreting antibody of predefined specificity. Nature 256:495–497

Kondo M, Weissman IL, Akashi K 1997 Identification of clonogenic common lymphoid progenitors in mouse bone marrow. Cell 91:661–672

Kondo M, Wagers AJ, Manz MG et al 2003 Biology of hematopoietic stem cells and progenitors: implications for clinical application. Annu Rev Immunol 21:759–806

Lagasse E, Weissman IL 1994 bcl-2 inhibits apoptosis of neutrophils but not their engulfment by macrophages. J Exp Med 179:1047–1052

Lagasse E, Connors H, Al-Dhalimy M et al 2000 Purified hematopoietic stem cells can differentiate into hepatocytes in vivo. Nat Med 6:1229–1234

Langhans T 1868 Ueber Riesenzellen mit wandstandigen Kernen in Tuberkeln und die fibrose Forn des Tuberkels. Virchows Arch Pathol Anat 42:382–404

Lapidot T, Sirard C, Vormoor J et al 1994 A cell initiating human acute myeloid leukaemia after transplantation into SCID mice. Nature 367:645–648

Magli MC, Iscove NN, Odartchenko N 1982 Transient nature of early haematopoietic spleen colonies. Nature 295:527–259

Main JM, Prehn RT 1955 Successful skin homografts after the administration of high dosage x-iradiation and homologous bone marrow. J Natl Cancer Inst 15:1023–1029

Manz MG, Traver D, Miyamoto T, Weissman IL, Akashi K 2001 Dendritic cell potentials of early lymphoid and myeloid progenitors. Blood 97:3333–3341

Manz MG, Miyamoto T, Akashi K, Weissman IL 2002 Prospective isolation of human clonogenic common myeloid progenitors. Proc Natl Acad Sci USA 99:11872–11877

Manz MG, Akashi K, Weissman IL 2004 Biology of hematopoietic stem and progenitor cells. In: Blume KG, Forman SJ, Appelbaum FR (eds) Thomas' hematopoietic cell transplantation. Blackwell Publishing Ltd, Malden MA

McCune JM, Namikawa R, Kaneshima H, Shultz LD, Lieberman M, Weissman IL 1988 The SCID–hu mouse: murine model for the analysis of human hematolymphoid differentiation and function. Science 241:1632–1639

McKinney-Freeman SL, Jackson KA, Camargo FD, Ferrari G, Mavilo F, Goodell MA 2002 Muscle-derived hematopoietic stem cells are hematopoietic in origin. Proc Natl Acad Sci USA 99:1341–1346

Mebius RE, Miyamoto T, Christensen J et al 2001 The fetal liver counterpart of adult common lymphoid progenitors gives rise to all lymphoid lineages, CD45+ CD4+ CD3 – cells, as well as macrophages. J Immunol 166:6593–6601

Mezey E, Chandross KJ, Harta G, Maki RA, McKercher SR 2000 Turning blood into brain: cells bearing neuronal antigens generated in vivo from bone marrow. Science 290:1779–1782

Michallet M, Philip T, Philip I et al 2000 Transplantation with selected autologous peripheral blood CD34+Thy1+ hematopoietic stem cells (HSCs) in multiple myeloma: impact of HSC dose on engraftment, safety, and immune reconstitution. Exp Hematol 28:858–870

Miyamoto T, Weissman IL, Akashi K 2000 AML1/ETO-expressing nonleukemic stem cells in acute myelogenous leukemia with 8;21 chromosomal translocation. Proc Natl Acad Sci USA 97:7521–7526

Miyamoto T, Iwasaki H, Reizis B et al 2002 Myeloid or lymphoid promiscuity as a critical step in hematopoietic lineage commitment. Dev Cell 3:137–147

Morrison SJ, Weissman IL 1994 The long-term repopulating subset of hematopoietic stem cells is deterministic and isolatable by phenotype. Immunity 1:661–173

Morrison SJ, Powse KR, Ho P, Weissman IL 1996 Telomerase activity in hematopoietic cells is associated with self- renewal potential. Immunity 5:207–216

Morrison SJ, Wandycz AM, Hemmati HD et al 1997a Identification of a lineage of multipotent hematopoietic progenitors. Development 124:1929–1939

Morrison SJ, Wright DE, Weissman IL 1997b Cyclophosphamide/granulocyte colony-stimulating factor induces hematopoietic stem cells to proliferate prior to mobilization. Proc Natl Acad Sci USA 94:1908–1913

Morrison SJ, White PM, Zock C, Anderson DJ 1999 Prospective identification, isolation by flow cytometry, and in vivo self-renewal of multipotent mammalian neural crest stem cells. Cell 96:737–749

Morshead CM, Benveniste P, Iscove NN, van der Kooy D 2002 Hematopoietic competence is a rare property of neural stem cells that may depend on genetic and epigenetic alterations. Nat Med 8:268–273

Muller-Sieburg CE, Whitlock CA, Weissman IL 1986 Isolation of two early B lymphocyte progenitors from mouse marrow: a committed pre-pre-B cell and a clonogenic Thy-1-lo hematopoietic stem cell. Cell 44:653–662

Na Nakorn T, Traver D, Weissman IL, Akashi K 2002 Myeloerythroid-restricted progenitors are sufficient to confer radioprotection and the majority of day 8 CFU–S. J Clin Invest 109:1579–1585

Na Nakorn T, Miyamoto T, Weissman IL 2003 Characterization of mouse clonogenic megakaryocyte progenitors. Proc Natl Acad Sci USA 100:205–210

Negrin RS, Atkinson K, LeemhuisT et al 2000 Transplantation of highly purified CD34+Thy-1+ hematopoietic stem cells in patients with metastatic breast cancer. Biol Blood Marrow Transplant 6:262–271

Nelson W, Nusse R 2004 Convergence of Wnt, b-catenin and cadherin pathways. Science 303:1483–1487

Nowell P, Cole L, Habermyer J et al 1956 Growth and continued function of rat marrow cells in x-radiated mice. Cancer Res 16:258

Orlic D, Kajstura J, Chimenti S et al 2001a Bone marrow cells regenerate infarcted myocardium. Nature 410:701–705

Orlic D, Kajstura J, Chimenti S et al 2001b Mobilized bone marrow cells repair the infracted heart, improving function and survival. Proc Natl Acad Sci USA 98:10344–10349

Palmer TD, Takahashi J, Gage FH 1997 The adult rat hippocampus contains primordial neural stem cells. Mol Cell Neurosci 8:389–404

Park IK, Qian D, Kiel M et al 2003 Bmi-1 is required for maintenance of adult self-renewing hematopoietic stem cells. Nature 423:302–305

Pittenger MF, Mackay AM, Beck SC et al 1999 Multilineage potential of adult human mesenchymal stem cells. Science 284:143–147

Reya T, Morrison SJ, Clarke MF, Weissman IL et al 2001 Stem cells, cancer, and cancer stem cells. Nature 414:105–111

Reya T, Duncan AW, Allies L et al 2003 Regulation of hematopoietic stem cells in self-renewal by the Wnt signaling pathway. Nature 423:409–414

Reynolds BA, Weiss S 1992 Generation of neurons and astrocytes from isolated cells of the adult mammalian central nervous system. Science 255:1707–1710

Rohrer JW, Vasa K, Lynch RG 1977 Myeloma cell immunoglobulin expression during in vivo growth in diffusion chambers: evidence for repetitive cycles of differentiation. J Immunol 119:861–866

Sata M, Saiura A, Kunisato A et al 2002 Hematopoietic stem cells differentiate into vascular cells that participate in the pathogenesis of atherosclerosis. Nat Med 8:403–409

Sell S, Pierce GB 1994 Maturation arrest of stem cell differentiation is a common pathway for the cellular origin of teratocarcinomas and epithelial cancers. Lab Invest 70:6–22

Sherwood RI, Christensen JL, Weissman IL, Wagers AJ 2004 Determinants of skeletal muscle contributions from circulating cells, bone marrow cells, and hematopoietic stem cells. Stem Cells 22 22:1292–1304

Shizuru JA, Jerabek L, Edwards CT, Weissman IL 1996 Transplantation of purified hematopoietic stem cells: requirements for overcoming the barriers of allogeneic engraftment. Biol Blood Marrow Transplant 2:3–14 (Erratum in Biol Blood Marrow Transplant 3:50)

Shizuru JA, Weissman IL, Kernoff R, Masek M, Scheffold YC 2000 Purified hematopoietic stem cell grafts induce tolerance to alloantigens and can mediate positive and negative T cell selection. Proc Natl Acad Sci USA 97:9555–9560

Short B, Brouard N, Occhiodoro-Scott T, Ramakrishnan A, Simmons PJ 2003 Mesenchymal stem cells. Arch Med Res 34:565–571

Siminovitch L, McCulloch E, Till J 1963 The distribution of colony-forming cells among spleen colonies. J Cell Comp Physiol 62:327–336

Smith LG, Weissman IL, Heimfeld S 1991 Clonal analysis of hematopoietic stem-cell differentiation in vivo. Proc Natl Acad Sci USA 88:2788–2792

Spangrude GJ, Heimfeld S, Weissman IL 1988 Purification and characterization of mouse hematopoietic stem cells. Science 241:58–62

Stadtmauer EA, O'Neill A, Goldstein LJ et al 2000 Conventional-dose chemotherapy compared with high-dose chemotherapy plus autologous hematopoietic stem-cell transplantation for metastatic breast cancer. Philadelphia Bone Marrow Transplant Group. N Engl J Med 342:1069–1076

Stamm C, Westphal B, Kleine HD et al 2003 Autologous bone-marrow stem-cell transplantation for myocardial regeneration. Lancet 361:45–46

Stemple DL, Anderson DJ 1992 Isolation of a stem cell for neurons and glia from the mammalian neural crest. Cell 71:973–785

Strauer BE, Brehm M, Zeus T et al 2002 Repair of infarcted myocardium by autologous intracoronary mononuclear bone marrow cell transplantation in humans. Circulation 106: 1913–1918

Tamaki S, Eckert K, He D et al 2002 Engraftment of sorted/expanded human central nervous system stem cells from fetal brain. J Neurosci Res 69:976–986

Taylor G, Lehrer MS, Jensen PJ, Sun TT, Lavker RM 2000 Involvement of follicular stem cells in forming not only the follicle but also the epidermis. Cell 102:451–461

Terada M, Hamazaki T, Oka M et al 2002 Bone marrow cells adopt the phenotype of other cells by spontaneous cell fusion. Nature 416:542–545

Thomas ED 1999 A history of haemopoietic cell transplantation. Br J Haematol 105: 330–339

Till JE, McCulloch EA 1961 A direct measurement of the radiation sensitivity of normal mouse bone marrow cells. Radiat Res 14:1419–1430

Traver D, Akashi K, Weissman IL, Lagasse E 1998 Mice defective in two apoptosis pathways in the myeloid lineage develop acute myeloblastic leukemia. Immunity 9:47–57

Traver D, Akashi K, Manz M et al 2000 Development of CD8α-positive dendritic cells from a common myeloid progenitor. Science 290:2152–2154

Traver D, Miyamoto T, Christensen J, Iwasaki-Arai J, Akashi K, Weissman IL 2001 Fetal liver myelopoiesis occurs through distinct, prospectively isolatable progenitor subsets. Blood 98: 627–635

Tsukamoto A, Weissman IL, Chen B et al 1995 Phenotypic and functional analysis of hematopoietic stem cells in mouse and human. In: Levitt D, Mertelsmann R (eds) Hematopoietic stem cells: biology and therapeutic applications. Marcel Dekker, New York, p 85–124

Uchida N 1992 Characterization of mouse hematopoietic stem cells. PhD thesis

Uchida N, Weissman IL 1992 Searching for hematopoietic stem cells: evidence that Thy-1.1lo Lin- Sca-1+ cells are the only stem cells in C57BL/Ka-Thy-1.1 bone marrow. J Exp Med 175:175–184

Uchida N, Sutton RE, Friera AM et al 1998a HIV, but not murine leukemia virus, vectors mediate high efficiency gene transfer into freshly isolated Go /G1 human hematopoietic stem cells. Proc Natl Acad Sci USA 95:11939–11944

Uchida N, Fujisaki T, Eaves AC, Eaves CJ 1998b High doses of purified stem cells cause early hematopoietic recovery in syngeneic and allogeneic hosts. J Clin Invest 101:961–966

Uchida N, Buck DW, He D et al 2000 Direct isolation of human central nervous system stem cells. Proc Natl Acad Sci USA 97:14720–14725

Varnum-Finney B, Xu L, Brashem-Stein C et al 2000 Pluripotent, cytokine-dependent, hematopoietic stem cells are immortalized by constitutive Notch1 signaling. Nat Med 6:1278–1281

Vignery A 2000 Osteoclasts and giant cells: macrophage-macrophage fusion mechanism. Int J Exp Pathol 81:291–304

Vose JM, Bierman PJ, Lynch JC et al 2001 Transplantation of highly purified CD34+Thy-1+ hematopoietic stem cells in patients with recurrent indolent non-Hodgkin's lymphoma. Biol Blood Marrow Transplant 7:680–687

Wagers AJ, Sherwood RI, Christensen JL, Weissman IL 2002 Little evidence for developmental plasticity of adult hematopoietic stem cells. Science 297:2256–2259

Wang X, Willenbring H, Akkari I et al 2003 Cell fusion is the principal source of bone-marrow-derived hepatocytes. Nature 422:897–901

Weissman IL 1963 Specific 'conditioning' of the host, in Section VIII of 'Tissue transplantation' by Eichwald EJ. Adv Biol Med Phys 9:160–177

Weissman IL 2000a Translating stem and progenitor cell biology to the clinic: barriers and opportunities. Science 287:1442–1446

Weissman IL 2000b Stem cells: units of development, units of regeneration, and units in evolution. Cell 100:157–168

Weissman IL, Anderson DJ, Gage F 2001 Stem and progenitor cells: origins, phenotypes, lineage commitments, and transdifferentiation. Annu Rev Cell Dev Biol 17:387–403

Whitlock CA, Tidmarsh GF, Muller-Sieburg C, Weissman IL 1987 Bone marrow stromal cells with lymphopoietic activity express high levels of a pre-B neoplasia-associated molecule. Cell 48:1009–1021

Willert K, Brown JD, Darenberg E et al 2003 Wnt proteins are lipid-modified and can act as stem cell growth factors. Nature 423:448–452

Wood GS, Mueller C, Warnke RA, Weissman IL 1988 In situ localization of HuHF serine protease mRNA and cytotoxic cell-associated antigens in human dermatoses. Am J Pathol 133:218–225

Wu A, Till JE, Siminovitch L, McCulloch EA 1963 Cytological evidence for a relationship between normal hematopoietic colony-forming cells and cells of the lymphoid system. J Exp Med 127:455–467

Wright DE, Cheshier SH, Wagers AJ, Randall TD, Christensen JL, Weissman IL 2001a Cyclophosphamide/granulocyte colony-stimulating factor causes selective mobilization of bone marrow hematopoietic stem cells into the blood after M phase of the cell cycle. Blood 97:2278–2285

Wright DE, Wagers AJ, Gulati AP, Johnson FL, Weissman IL 2001b Physiological migration of hematopoietic stem and progenitor cells. Science 294:1933–1936

Ying QL, Nichols J, Evans EP, Smith AG 2002 Changing potency by spontaneous fusion. Nature 416:545–548

Yuan Y, Zhou L, Miyamoto T et al 2001 AML1-ETO expression is directly involved in the development of acute myeloid leukemia in the presence of additional mutations. Proc Natl Acad Sci USA 98:10398–10403

DISCUSSION

Gearhart: You commented earlier on what you thought was going on in George Daley's embryonic stem (ES) cell work. Have you seen any convincing evidence that it will be possible to take ES cells and derive HSCs from them?

Weissman: We isolate cells that have the phenotype of HSCs 4–6 days after we remove Lef. It works best if they are on Op9 or Ad6 monolayers. They certainly have the characteristics of HSCs and if we re-transfer them into the stromal cell assays for stem cells they work fine, but they don't transplant *in vivo*. This is why I think they are like the yolk sac stage of development. When we Wnt stimulate purified mouse HSCs, HoxB4 and Notch 1 are up-regulated, and there is good independent evidence that they are in the pathway of proliferation, expansion or maturation. How this works in the system is unknown. This gives us hope that if we do develop HSCs from ES cells, sooner or later you are going to find a reductionist way to expand them and use them. For the brain transplant experiments that were done, it may be that it is more straightforward in neural pathways to get early neural stem cells that can be used. Each and every tissue is a different case.

Wilmut: Would the treatment you need to give the patient to destroy their endogenous haematopoietic system be lethal in the absence of effective replacement cells?

Weissman: No, you can give a sublethal dose of irradiation now. Many of the leukaemia cell bone marrow transplant (BMT) doctors give what is called a non-myeloablative regimen of 200–300 rads, which is close to 500–600 rads in a mouse. This wipes out much but not all of the haematopoietic system. They then give Fludarabine, which in mice we can mimic with antibodies to T cells and NK cells. This is therefore a non-myeloablative but lymphoablative protocol.

Wilmut: Does this not do any damage to other tissue?

Weissman: It doesn't do as much damage as having that disease.

Wilmut: It is a step from doing this with an organ transplant to doing it with diabetes.

Weissman: If you are doing an organ transplant by the current method you are putting the patient through much more than you would in this. They are on constant immunosuppression, they get infections, and quite a few get EBV lymphomas and other tumours.

Wilmut: You have started with diabetes, but you have obviously thought through other autoimmune diseases where the condition has already arisen. Which other diseases would you hope to treat in this way?

Weissman: Multiple sclerosis (MS) would be next. We know from our nervous system precursors that in a genetically demyelinated mouse, Shiverer, human neural stem cells replace the mouse oligodendrocytes that are attached to the axons and give the full 16 wraps around the axons.

Soria: Are you talking about autologous transplantation of HSCs or allogeneic transplant?

Weissman: I'm talking about allogeneic transplant of HSCs either to induce transplant tolerance or block autoimmunity. The mechanism by which an allogeneic haematopoietic system that itself has no genetic risk factors for the autoimmune disease stops the autoimmune disease that is already ongoing is still not understood. I just know it is effective.

Soria: There are some reports in which establishment of mixed hematopoietic chimerisms resulted in the induction of transplantation tolerance to any other organ from the same donor (Cosimi & Sachs 2004, Buhler et al 2002).

Weissman: The problem with this and similar studies is that they didn't go far enough in the myeloablation to get a good donor chimera. Any of you who have dealt with clinical groups will know that they are not coming at this from the same viewpoint as researchers. Their first issue is to do the least potential harm. This is what they are supposed to do. Getting a real trial will take a lot of convincing animal experiments.

Solter: I can sort of see why there are these observed differences in plasticity between different labs. You mentioned specifically the NOD mice in which some people do see islet regeneration after injection of bone marrow cells (Kodama et al 2003), while you don't. In this case you do exactly the same thing, so I can't explain it.

Weissman: Nor can I. Perhaps they are better clinicians at keeping the animals alive after the BMT, or maybe they caught it before it was a full depletion. I don't know. I don't understand why we don't game the same data in so many cases when we try to repeat experiments. Our paper on the hearts was rejected by all the reviewers because we couldn't explain why we couldn't repeat another finding. I said we just couldn't explain it, and pointed out that there are now five clinical trials already published in bone marrow injection into the heart of patients with coronaries. I said to the journal editor that they should be responsible enough to publish the other point of view.

Nadal-Ginard: This has been repeated in a range of animals by a number of groups. The fact that I cannot sing like Pavarotti doesn't mean that Pavarotti cannot sing. In terms of the heart, the mouse transplants are heterotopic transplants. Have you tried to do this in a real transplant where the heart has to work, and see whether you can maintain function?

Weissman: A long time ago we did heterotopic heart transplants, but this is tough in mice.

Goldman: You mentioned MS as a target. What kind of patient selection do you have in mind?

Weissman: We need to know whether in late-stage MS patients have a continuing autoimmune problem. When I talk to people they say that it is all burned out, but how do they know? It is not really clear.

Goldman: By the time patients get to that point they may have little myelin left and not many axons. But what would you do about the patient selection?

Weissman: What would I do? I would call Anders Björklund or Steve Goldman and ask them!

Grompe: With regards to reducing tolerance with bone marrow transplantation, even sublethal radiation is not innocuous. It is the equivalent of being exposed to the Hiroshima atom bomb.

Weissman: It depends how far away from the bomb!

Grompe: In terms of diabetes, for example, is this something that one would consider worth risking?

Weissman: Your institution review boards would help you make that decision. Every one of my friends who had juvenile diabetes when we were growing up is long dead. They had pretty good insulin coverage. As you know, the latest trial on insulin pumps has shown that the secondary complications of diabetes are still not mitigated. It is a social as well as institutional issue.

Nadal-Ginard: So bone marrow has stem cells. We all agree on this. What other role do the stem cells have? Specifically, what are the consequences of whole body irradiation for all those non-haematopoietic marrow-derived stem cells?

Weissman: There is a lot of experience with people who get the equivalent of whole body radiation. It would be the same protocol as you use now. What are the consequences of haematopoietic transplants, autologous or allogeneic, over the long term? There are some. But I don't think we see massive effects, for example, in the CNS. If you look at Arturo Alvarez-Buylla's experiments with radiation (I think Fred Gage has done this also), the dose of irradiation that is sufficient to kill a mouse by haematopoietic failure does not get rid of HSCs. It gets rid of enough that you are in trouble. In fact, I don't have any transplants in an allogeneic system where I have used stem cells where there isn't host haematopoiesis living with the donor haematopoiesis. You can do straight radiobiology. The dose to kill 90% of the cells by mitotic death is 300 rads. Once you determine the number of stem cells and the dose you are going to give for radiation, you can calculate pretty directly how many HSCs are left. We went through this in a very formal fashion for HSCs. One would have to look at brain, skin and other tissues in the same way. These aren't usually lifelong problems with patients who have received myeloablative therapies.

Soria: Is irradiation the only way to do this?

Weissman: No, it can also be done with combination chemotherapies. This is probably what would be done in humans because this is what we have most experience in.

Soria: In this case, it will be closer to clinical assays.

Weissman: I think you would end up giving a submyeloablative dose. The group in Seattle gives 200 rads plus some chemotherapy. They are going to lead the way to find what the non-myeloablative requirement is.

Goldman: With regards to brain, the sublethal radiotherapy doses used in children have been associated with late dementias. Theo Palmer has pointed out that a lot of this may be through radiotherapy changing the hippocampal environment enough to suppress hippocampal neurogenesis.

Weissman: We need to look carefully at children, because their brains are still growing. I am not a snake oil salesman: I am not going to say that we are going to treat all diseases. But we will find candidates in every one of those autoimmune diseases where you can do something.

Grompe: The reason that the brain is included in irradiation for paediatric bone marrow transplantion is because the indication is usually leukaemia. For haematopoietic engraftment you wouldn't need to irradiate the head.

Weissman: No, you need to because there is bone marrow and there are lymphoid cells there.

Grompe: It would probably be possible to get substantial haematopoietic engraftment and spare the brain from irradiation.

Nadal-Ginard: There is a high incidence of heart failure in these patients. It isn't possible to protect the heart.

References

Buhler LH, Spitzer TR, Sykes M et al 2002 Induction of kidney allograft tolerance after transient lymphohaematopoietic chimerism in patients with multiple myeloma and end-stage renal disease. Transplantation 74:1405–1409

Cosimi AB, Sachs DH 2004 Mixed chimierism and transplantation tolerance. Transplantation 77:943–946

Kodama S, Kühtreiber W, Fujimura S, Dale EA, Faustman DL 2003 Islet regeneration during reversal of autoimmune diabetes in NOD mice. Science 302:1223–1227

Multipotent adult progenitor cells: an update

Catherine M. Verfaillie

Department of Medicine, Stem Cell Institute, University of Minnesota, Moos Tower 14-287A, 420 Delaware Street, Minneapolis, MN 55455, USA

Abstract. The quintessential stem cell is the embryonal stem (ES) cell which has unlimited self-renewal and multipotent differentiation potential. Stem cells have also been identified in most tissues. Compared with ES cells, tissue-specific stem cells have less self-renewal ability and, although they differentiate into multiple lineages, they are not multipotent. A large number of recent published studies have suggested that tissue-specific stem cells may have the ability to generate cells of tissues from unrelated organs. We have identified a population of primitive cells in normal human, rodent, and possibly other mammalian postnatal tissues that have, at the single cell level, multipotent differentiation and extensive proliferation potential, which we named multipotent adult progenitor cell (MAPC). We will discuss studies aimed at determining whether MAPCs also exist in bone marrow of other species, whether MAPCs exist *in vivo* or are a culture phenomenon, to further characterize MAPCs using gene expression profiling, and effects of *in vivo* transplantation of MAPCs.

2005 Stem cells: nuclear reprogramming and therapeutic applications. Wiley, Chichester (Novartis Foundation Symposium 265) p 55–65

The quintessential stem cell is the embryonal stem (ES) cell which has unlimited self-renewal and multipotent differentiation potential. Stem cells have also been identified in most tissues. Compared with ES cells, tissue specific stem cells have less self-renewal ability and, although they differentiate into multiple lineages, they are not multipotent. A large number of recently published studies have suggested that tissue-specific stem cells may have the ability to generate cells of tissues from unrelated organs. We have identified a population of primitive cells in normal human, rodent, and possibly other mammalian post-natal tissues that have, at the single cell level, multipotent differentiation and extensive proliferation potential, which we named multipotent adult progenitor cells (MAPCs).

At the meeting updated information was provided regarding the possible aetiology of the cell, i.e. is the cell a culture artefact or does the cell exist as such *in vivo*. In addition, updated information was provided regarding the potential differentiation ability of MAPCs.

Phenotype of MAPCs

MAPCs can be cultured from human, mouse and rat bone marrow (BM). Unlike MSC, MAPCs do not express major histocompatibiliy (MHC) class I antigens, do not express or express only low levels of the CD44 antigen, and are CD105 (also endoglin, or SH2) negative (Jiang et al 2002a, Reyes et al 2001). Unlike haematopoietic stem cells (HSCs), MAPCs do not express CD45, CD34, and c-Kit (Jiang et al 2002a, Reyes et al 2001), but like HSCs, MAPCs express Thy1, AC133 (human MAPC) and Sca1 (mouse), albeit at low levels (Jiang et al 2002a, Reyes et al 2001). In the mouse, MAPCs express low levels of stage-specific embryonic antigen (SSEA)-1, and express low levels of the transcription factors Oct4 (Oct3 in human and monkey), nanog and Rex1, known to be important for maintaining embryonic stem (ES) cells undifferentiated (Niwa et al 2000) and to be down-regulated when ES cells undergo somatic cell commitment and differentiation (Jiang et al 2002a).

MAPCs can also be isolated from other tissues, and other species

We also showed that MAPCs can be cultured from mouse brain and mouse muscle (Jiang et al 2002b). Of note, the differentiation potential and expressed gene profile of MAPCs derived from the different tissues appears to be highly similar. These studies used whole brain and muscle tissue as the initiating cell population, therefore containing more than neural cells and muscle cells, respectively. The implications of this will be discussed below. Studies are ongoing to determine if cultivation of MAPC from other organs is possible, and whether culture of MAPC, like ES cells, is mouse-strain dependent. Initial studies suggest that a population of MAPC-like cells can also be cultured from bone marrow from cynomologous monkeys (unpublished observations; studies done by our collaborator Felipe Prosper, University of Navarra, Pamplona, Spain) and from bone marrow of swine (unpublished observations; studies done at the University of Minnesota).

Non-senescent nature of MAPCs

Unlike most adult somatic stem cells, MAPCs proliferate without obvious signs of senescence, and have active telomerase. In humans, the length of MAPC telomeres is 3–5 kB longer than in neutrophils and lymphocytes, and telomere length is not different when MAPCs are derived from young or old donors (Reyes et al 2001). This suggests that MAPCs are derived from a population of cells that either has active telomerase *in vivo*, or that is highly quiescent *in vivo*, and therefore have not yet incurred telomere shortening *in vivo*. In human MAPC cultures we have not yet seen cytogenetic abnormalities. As human MAPCs are, however, undergoing symmetrical cell divisions, it remains possible that despite the lack of gross cytogenetic changes, minor mutations accumulate over time. We are therefore planning to use comparative genomic hybridization (CGH) to address the question at what time genetic abnormalities occur, if they do.

However, several subpopulations of mouse MAPCs, and to a lesser extent rat MAPCs, have become aneuploid, tetraploid or hypodiploid, even though additional subpopulations thawed subsequently were cytogenetically normal. Cytogenetic abnormalities are seen more frequently once mouse (and rat) MAPC have been expanded for >60–70 population doublings (PDs) and following repeated cryopreservations and thawing episodes. This characteristic of mouse MAPCs is not dissimilar from other mouse cell populations, including mouse ES cells. We are currently testing the influence of serum, cytokines, and ambient O_2 levels on this phenomenon.

Stringent culture conditions required for maintenance of the undifferentiated state of MAPC

Culture of MAPCs is, however, technically demanding. Major factors that play a role in successful maintenance of MAPC include cell density, CO_2 and O_2 concentration, and the lot of fetal calf serum that is used. Control of cell density appears to be species specific: mouse and rat MAPCs need to be maintained at densities between 200 and 800 cells/cm^2, swine and cynomologous monkey MAPCs between 500 and 2000 cells/cm^2, whereas human and perhaps dog MAPCs need to be maintained between 1500 and 3000 cells/cm^2. The reason why MAPCs tend to differentiate when maintained at higher densities is not known. However, for MAPCs to have clinical relevance, this will need to be overcome. Gene array and proteomics studies are ongoing to identify the contact and/or soluble factors that may be responsible for causing differentiation when MAPCs are maintained at higher densities.

In vitro differentiation potential of MAPCs

We published last year that human, mouse and rat MAPCs can be successfully differentiated into typical mesenchymal lineage cells, including osteoblasts, chondroblasts, adipocytes and skeletal myoblasts (Reyes et al 2001). In addition, human, mouse and rat MAPCs can be induced to differentiate into cells with morphological, phenotypic and functional characteristics of endothelial cells (Reyes et al 2002) and hepatocytes (Schwartz et al 2002).

Possible mechanisms underlying the phenomenon of multipotent adult progenitor cells

Currently we do not fully understand the mechanism(s) underlying the culture selection of MAPCs. We have definitive data to demonstrate that the pluripotency of MAPCs is not due to co-culture of several stem cells.

Pluripotency cannot be attributed to multiple stem cells

First, using retroviral marking studies we have definitively proved that a single cell can differentiate *in vitro* to cells of mesoderm, both mesenchymal and non-mesenchymal, neuroectoderm and hepatocyte-like cells, and this applies for human (Reyes et al 2001, Schwartz et al 2002), mouse and rat MAPCs (Jiang et al 2002a, Schwartz et al 2002). Second, we have shown that a single mouse MAPC is sufficient for generation of chimeric animals (Jiang et al 2002a). Indeed, we published that one-third of animals born from blastocysts in which a single MAPC was injected were chimeric with chimerism degrees varying between 1 and 45%. Therefore, this rules out the possibility that the pluripotent nature of these cells is due to the co-existence in culture of multiple somatic stem cells.

Cell fusion is not a likely explanation

A second possibility for the greater degree of differentiation potential would be that cells undergo fusion and acquire via this mechanism greater pluripotency. Fusion has been shown to be responsible for apparent ES characteristics of marrow and neural stem cells (Terada et al 2002, Ying et al 2002, Lagasse et al 2000, Wang et al 2003). We do not believe that this phenomenon underlies the observation that MAPCs are pluripotent. Cultivation and differentiation *in vitro* (in general, except the final differentiation step for neuroectoderm) does not require that MAPCs are co-cultured with other cells, making the likelihood that MAPCs are the result of fusion very low. Smith and colleagues suggested in a recent commentary that MAPCs could be caused by fusion of multiple cell types early on during culture leading to reprogramming of the genetic information and pluripotency. However, we have no evidence that hMAPCs or rodent MAPCs are tetraploid or aneuploid early during culture, making this possibility less likely. Nevertheless, studies are ongoing to rule this out. Our *in vivo* studies were not set up to fully be capable of ruling out this possibility. However, a number of findings suggest that fusion may not likely be the cause for the engraftment seen postnatally, nor the chimerism in the blastocyst injection experiment. The frequency of the fusion event described for the ES-BM, ES-NSC, and HSC-hepatocyte fusion was in general very low, i.e. 1/100 000 cells. Expansion of such fused cells could only be detected when drug selection was applied in the *in vitro* systems, and withdrawal of

NTBC (2-[2-nitro-4-trifluoro-methylbenzoyl]-1,3-cyclohexanedione) in the *Fah* mouse model was used to select for cells expressing the *Fah* gene. The percent engraftment seen in our post-natal transplant models was in the range of 1–9%. The frequency of chimerism seen in blastocyst injection studies ranged from 33% to 80% when 1 and 10–12 MAPCs were injected, respectively. These frequencies are significantly higher than what has been described for fusion events with ES cells *in vitro*, and in the haematopoietic stem cell (HSC)–hepatocyte fusion studies *in vivo*. Furthermore, in contrast to what was described in the papers indicating that fusion may be responsible for apparent plasticity, all *in vivo* studies done with MAPCs were done without selectable pressure, mainly in non-injured animals. Therefore, it is less likely that the pluripotent behaviour of MAPCs *in vivo* is due to fusion between the MAPCs and the tissues where they engraft/contribute to. Specific studies using Z/EG mice and cre-recombinase mice are currently being designed to formally rule this out.

Primitive ES-like cells that persists versus dedifferentiation

Currently, we do not have proof that MAPCs exist as such *in vivo*. Until we have positive selectable markers for MAPCs, this question will be difficult to answer. If the cell exists *in vivo*, one might hypothesize that it is derived for instance from primordial germ cells that migrated aberrantly to tissues outside the gonads during development. It is, however, also possible that removal of certain (stem) cells from their *in vivo* environment results in 'reprogramming' of the cell to acquire greater pluripotency. The studies on human MAPCs suggest that such a cell that might undergo a degree of reprogramming and is likely a protected (stem) cell *in vivo*, as telomere length of MAPCs from younger and older donors is similar, and significantly longer than what is found in haematopoietic cells from the same donor. The fact that MAPCs can be isolated from multiple tissues might argue that stem cells from each tissue might be able to be reprogrammed. However, as was indicated above, the studies in which different organs were used as the initiating cell population for generation of MAPCs did not purify tissue-specific cells or stem cells. Therefore, an alternative explanation is that the same cells isolated from bone marrow that can give rise to MAPCs in culture might circulate, and be collected from other organs. However, we have until now been unsuccessful in isolating MAPCs from blood or from umbilical cord blood, arguing against this phenomenon. Finally, cells selected from the different organs could be the same cells resident in multiple organs, such as MSCs that are present in different locations, or cells associated with tissues present in all organs such as for instance blood vessels. Studies are ongoing to determine which of these many possibilities is correct.

Conclusion

We believe that MAPCs would have clinical relevance whether they exist *in vivo*, or are created *in vitro*. However, understanding the nature of the cell will have impact on how one would approach their clinical use. If they exist *in vivo*, it will be important to learn where they are located, and to determine whether their migration, expansion and differentiation in a tissue specific manner can be induced and controlled *in vivo*. If they are a culture creation, understanding the mechanism underlying the reprogramming event will be important as that might allow this phenomenon to happen on a more routine and controlled basis.

Either way, a long road lies ahead before MAPCs might be applicable in clinical trials. Hurdles to be overcome include development of robust culture systems that will allow automization. For other stem cells, including ES cells, we will need to determine in preclinical models whether undifferentiated vs. lineage-committed vs. terminally differentiated cells should be used to treat a variety of disorders. If lineage-committed or terminally differentiated cells will be needed, robust clinical-scale differentiation cultures will need to be developed. Furthermore, studies will need to be performed to demonstrate whether potentially contaminating undifferentiated MAPCs will interfere with engraftment, and/or differentiate inappropriately *in vivo*. Likewise, studies aimed at determining what level of HLA-mismatch will be tolerated in transplantations, whether tolerization via haematopoietic engraftment from MAPCs will be required. As is also the case for other extensively cultured cells, we will need to further determine whether prolonged expansion leads to genetic abnormalities in cells that might lead to malignancies when transplanted *in vivo*.

References

Jiang Y, Jahagirdar B, Reyes M et al 2002a Pluripotent nature of adult marrow derived mesenchymal stem cells. Nature 418:41–49

Jiang Y, Vaessen B, Lenvik T, Blackstad M, Reyes M, Verfaillie CM 2002b Multipotent progenitor cells can be isolated from post-natal murine bone marrow, muscle and brain. Exp Hematol 30:896–204

Lagasse E, Connors H, Al-Dhalimy M et al 2000 Purified hematopoietic stem cells can differentiate into hepatocytes in vivo. Nat Med 6:1229–1234

Niwa H, Miyazaki J, Smith AG 2000 Quantitative expression of Oct-3/4 defines differentiation, dedifferentiation or self-renewal of ES cells. Nat Genet 24:372–376

Reyes M, Lund T, Lenvik T, Aguiar D, Koodie L, Verfaillie CM 2001 Purification and ex vivo expansion of postnatal human marrow mesodermal progenitor cells. Blood 98:2615–2625

Reyes M, Dudek A, Jahagirdar B, Koodie K, Marker PH, Verfaillie CM 2002 Origin of endothelial progenitors in human post-natal bone marrow. J Clin Invest 109:337–346

Schwartz RE, Reyes M, Koodie L et al 2002 Multipotent adult progenitor cells from bone marrow differentiate into functional hepatocyte-like cells. J Clin Invest 96:1291–1302

Terada N, Hamazaki T, Oka M et al 2002 Bone marrow cells adopt the phenotype of other cells by spontaneous cell fusion. Nature 416:542–545

Wang X, Willenbring H, Akkari Y et al 2003 Cell fusion is the principal source of bone-marrow-derived hepatocytes. Nature 422:897–901
Ying QY, Nichols J, Evans EP, Smith AG 2002 Changing potency by spontaneous fusion. Nature 416:545–548

DISCUSSION

Gearhart: Have you got the germ cells from your chimeras?

Verfaillie: We have male cell lines now, so we are in the process of injecting them. All our initial cell lines were female for some reason. To make a cell line takes about 5–6 months.

Gearhart: Have you done somatic cell nuclear transfer (SCNT) on the MAPCs, from the standpoint of looking at the nuclei to see whether the frequency of generating an embryo is high relative to somatic lines?

Verfaillie: No.

Gepstein: With regard to the ES cells, in *invivo* experiments it seems that the adult tissues don't provide a nice environment for differentiation into the appropriate tissue. Did you also transplant cells that were differentiated *ex vivo* into the same organ? What happens to them?

Verfaillie: We have not done this yet. In the neuronal differentiation system we intend to put in cells that have nestin expressed at high levels. We want to see whether this gives a better degree of engraftment. With markers we are planning to put in undifferentiated cells as well as cells that are at certain points along the differentiation pattern.

Nadal-Ginard: What is the frequency of your cells in bone marrow?

Verfaillie: We did the purimycin selection and in 4–5 week old mice it is probably one in a million. In newborn mice it is likely 1 in 100 000. For other tissues it is probably much lower than bone marrow, even though we have been able to grow them out from brain and muscle. We haven't gone prenatal: the problem is that if we use the *Oct4* selection we will run into the primordial germ cells if we start going very early on. There is no easy way to discriminate between these two.

Björklund: You mentioned that you have studied neuroepithelial differentiation after stroke. Can you tell us a little more about how this experiment was done?

Verfaillie: That wasn't a stroke; it was an intact newborn animal, and we did intraventricular injection in newborns.

Björklund: Had these cells been predifferentiated?

Verfaillie: No, they were undifferentiated cells.

Björklund: What happened to the cells that didn't integrate? Did you find any clumps or tumour formation, or any sign that non-differentiated cells survived?

Verfaillie: We see a lot of cells sitting around blood vessels. These cells haven't differentiated and don't express any neural markers, as far as we can tell.

Björklund: But you don't see anything that hasn't integrated into the ventricle.

Verfaillie: There are some cells still on the inside of the ventricle, and they don't form tumours as far as we can tell. This is just with four or five different animals.

Björklund: Do you have any sense of what proportion of the total number of cells ended up with a differentiated morphology?

Verfaillie: I can't answer that at this point.

Björklund: It is interesting to compare your results with those obtained with ES cells. If ES cells are injected into the brain parenchyma they grow as clumps. These clumps grow to a variable extent. If you don't predifferentiate the cells or, enrich or sort them in some way, there is variable production of teratoma-like tumours. In these clumps you may have cells that differentiate, but they migrate very poorly into the host tissue. Your cells, by contrast, appear to be much more migratory. Can you generate neurospheres from them?

Verfaillie: We are just starting to do this. On day 4 of our neural differentiation scheme we get nestin to form and some of the early transcription factors. We are now taking them off the dish to see whether we can make them into neural spheres. So far we don't know whether we can or not.

Weissman: In the intraventricular experiment it looked like you had extensive migration and differentiation at 3 and 4 weeks. I don't think we get as much with human neurospheres at that time. We have to wait many weeks longer to see that.

Björklund: It is also a small fraction of the implanted cells that actually integrates into the host parenchyma.

Verfaillie: We use about 300 000 cells per injection. This was rat into rat, and this is a species that develops quicker than human.

Björklund: In the intraventricular experiment, do you have any idea of how neurogenic cells compare to non-neurogenic cells?

Verfaillie: We have injected them in multiple different locations, but we don't have the data yet.

Goldman: What was the exact age of injection?

Verfaillie: The first week: day 3.

Goldman: If you put neural stem cells in at that stage, you won't see neurons produced except in persistently neurogenic regions. Most of the donor cells, especially the long distance migrants, will become oligodendrocytes, and those that don't will become astrocytes. I was struck by the neuronal phenotype assumed by your passaged cells, because you wouldn't have even expected that from neural progenitors or neural stem cells. Do you have any hint about potential differentiation of your cells prior to implantation?

Verfaillie: I can't tell you that they are 100% homogeneous MAPC cells. Usually the good cultures are 80–90% *Oct4* positive, which would mean that up to about 30% are not.

Björklund: Are they all nestin negative?

Verfaillie: They display very low levels of nestin.

Björklund: This suggests that they are more or less homogenous and very poorly differentiated. Have you looked at *Sox1* and *Sox2* expression?

Verfaillie: We find it at extremely low levels. All of these come on during the first week of differentiation *in vitro*.

Goldman: Have you sorted *Oct4*-GFP cells directly from your tissue samples?

Verfaillie: No, we haven't. We are now looking at the newborn animals to see how frequent this is. In the adult mice this would be extremely hard to sort out because the frequency is less than one in a million.

Nadal-Ginard: Apart from the MAPCs, can you get cell lines that are constricted?

Verfaillie: We have cell lines that, if induced to differentiate, don't make certain cell lineages. We have some that are very good at making endoderm liver-like cells but can't make the other cells.

Grompe: I wanted to make a comment regarding the increasingly popular Cre-Lox approach to fusion. When there is a positive signal in the transplantation, you can be reasonably sure that fusion has occurred. But the other way round, it is not a solid tool to rule out fusion because it is dependent on the expression of two transgenes in the same cell. As we now know, some of these fusion products can rapidly undergo reduction divisions, so the transgenes could be segregated from each other. I think the gold standard is to use an expression-independent marker. Cytogenetics works. We know that transgenes can be quite mosaic in their expression. The positive controls people mention are breeding the Cre and LoxP together and getting loss of blue or green cells, but this is a very different situation because the Cre and LoxP have been together in the same cells since conception. I would argue that it is a good way to look positively for fusion but it is not a good way to rule out fusion.

Soria: Are MAPCs present in the embryo or the umbilical cord?

Verfaillie: We haven't looked before birth. We have tried three times to work from human umbilical cord blood and have failed, but this doesn't mean that they don't exist.

Wilmut: In earlier presentations I have heard you say that the growth of the cells changed by 30 passages. Do you still see this, and across all species?

Verfaillie: Pretty much. With human they seem to grow quicker for the first 30–40 passages and then slow down a bit. The same is true in mouse, although it is harder to say when because we did the selection differently.

Gurdon: When you isolate cells from the brain or the muscle, do you at first see none of the cell type-specific markers expressed?

Verfaillie: By RT-PCR we don't find the ES-like genes in the bulk population. What we have in the culture system is a mixed bag and this is a difficult question to answer.

Gearhart: To clarify: is the relationship between MSCs and MAPCs that one is derived from the other? Is this clear, or is it an issue of a heterogeneous population to begin with? What is your best guess on this? Are there other distinguishing features between the two populations?

Verfaillie: I don't know the answer: this probably has to come from trying to isolate them on day 0 as pure MSCs or MAPCs, if such cells exist on day 0. MAPCs can make all mesenchymal lineages, so they would be the precursors for MSCs. But are they a distinct population, or is it like a haematopoietic hierarchy with a continuum of cells that have more or less potential? I don't know. The ones we have in culture are phenotypcially distinct; and looking by gene array, for what this is worth, the difference between human MSCs and MAPCs is about 2000–4000 genes if you use a criterion of two to threefold expression difference. This would suggest that their molecular signature is different, although this could be a culturally induced phenomenon.

Gearhart: Is this comparing cells isolated from the same source?

Verfaillie: This is looking at multiple MAPC and MSC populations.

Soria: In some of the experiments the cells were transplanted without any differentiation protocol. Once they have been transplanted do they stop dividing?

Verfaillie: We have been doing BrDU staining to figure out whether they stop dividing or not. With the intraventricular injection experiment, I don't know whether the MAPCs divided or whether what they created divided, because there are more cells with the donor marker in these mice than we introduced. Something has divided.

Soria: I wonder if the implant location that controls cell proliferation, for example, is intravenous. However, when mouse non-differentiated ES cells are transplanted into the spleen they form teratomas (Soria et al 2000).

Verfaillie: So far we have not seen teratomas. I don't know whether we did the right experiment or not, but so far we have not seen them.

Soria: Have you any idea about the physiological role of these cells?

Verfaillie: That question assumes they exist, which we don't know for sure. If they do, I don't know their role. We don't fix up cells all the time: we die from multiple diseases. If they do exist, we don't know that they would become activated, or whether they move. We are hoping to use the same parabiotic model that Irv Weissman described. The problem with those mice, though, is that they need to be aged, and so the number of cells goes down making it a hard experiment to do. We are in the process of setting this up, and these experiments would address the physiological relevance of these cells.

Nadal-Ginard: Can you isolate MAPCs from peripheral blood?

Verfaillie: So far we have not been able to do this. If the frequency is one in a million in the bone marrow, it might be quite a bit lower in the peripheral blood.

Grompe: The transplantation experiments would suggest that they are not normally circulating cells. You don't find them circling around in the peripheral blood of transplanted mice like you do HSCs.

Verfaillie: We have not looked at whether we can grow them back if we were to draw blood 48 h after injection, for example. We also have no idea of the mechanisms of homing of these cells, and why they go to some organs and not others.

Reference

Soria B, Roche E, Berna G, Leon-Quinto T, Reig JA, Martin F 2000 Insulin-secreting cells derived from embryonic stem cells normalize glycemia in streptozotocin-induced diabetic mice. Diabetes 49:157–162

Neural progenitor cells of the adult brain

Steven A. Goldman and Fraser Sim

Department of Neurology, University of Rochester Medical Center, 601 Elmwood Avenue, MRBX, Box 645, Rochester, NY 14642, USA

Abstract. Recent studies have substantially expanded our conception of the types of progenitor cells that continue to reside in the adult nervous system, and their respective roles in the normal maintenance of the brain and spinal cord. In the adult, neural stem cells persist within the forebrain ventricular zone, and give rise to a variety of more restricted progenitor phenotypes. The major progenitor pools of the adult human brain, each of which has now been isolated to purity, include ventricular zone neuronal progenitor cells, hippocampal neuronal progenitors and parenchymal glial progenitor cells. Each of these phenotypes exists within a local environmental niche, which tightly regulates both the mitotic activity and derivatives of its resident progenitors. Within these niches, both neuronal and glial progenitor cells may reside as transit amplifying pools, by which lineage-biased progenitors expand to replenish discrete mature phenotypes. The largest such pool appears to be that of the parenchymal glial progenitor cell. These cells are restricted to glial phenotype by their local white matter environment, but upon removal from that setting, can expand to generate neurons as well as glia. Thus, at least some populations of adult glial progenitors retain both multilineage capacity and mitotic competence, and as such may be able to regenerate the tissue-specified stem cells from which they derived. This presentation will focus on the interactions of the adult human white matter with its resident glial progenitor cells, emphasizing those pathways that regulate the maintenance of these cells in their undifferentiated state, in particular those that may be shared with ventricular zone neural stem cells.

2005 Stem cells: nuclear reprogramming and therapeutic applications. Wiley, Chichester (Novartis Foundation Symposium 265) p 66–91

Neural stem and progenitor cells are dispersed widely throughout the adult vertebrate brain

Gone are the days when the adult CNS was considered a structurally static organ. Studies in the 1970s and 1980s in teleost fish, songbirds and rodents revealed the persistence of neurogenesis in a variety of regions of the CNS. On the basis of these comparative studies, neurogenesis within the adult rodent brain was intensively studied by a number of groups in the early 1990s. These studies revealed that distinct populations of both neuronally restricted progenitors and multipotential neural stem cells persist within the ventricular zone of the adult brain. The cells appear to reside largely within the subependymal layer of the ventricular wall, within which they may represent residual elements of the fetal neuroepithelium (Weiss et al 1996, Morshead et al 1994, Kirschenbaum & Goldman 1995). Neural stem cells, defined by their self-renewal capacity as well as by their multilineage competence, seem to be especially abundant in the forebrain, in which they have been identified in species ranging from mice to human (Goldman 1998, 2003, Gage 2000). In addition, phenotypically committed neuronal progenitor cells remain widely distributed in the ventricular wall (Kirschenbaum & Goldman 1995), throughout both its rostral extensions to the olfactory bulb, and caudally through the subgranular zone to the dentate gyrus of the hippocampus (Doetsch & Alvarez-Buylla 1996, Doetsch et al 1997, Palmer et al 1997, Luskin 1993, Kornack & Rakic 1999). Besides these neurogenic progenitor cell populations, glial progenitors also reside within both the ventricular zone and tissue parenchyma (Roy et al 1999). The latter may be a heterogeneous group, comprising both bipotential astrocyte–oligodendrocyte progenitors as well as even less-restricted multipotential progenitors, biased to gliogenesis but nonetheless still capable of neurogenesis. Together, these different classes of progenitors constitute the major known categories of neural precursor cells in the adult CNS (reviewed in Alvarez-Buylla & Garcia-Verdugo 2002, Goldman 2001, 2003).

Progenitor cells are abundant in the adult human brain

Humans harbour both neuronal progenitor cells and neural stem cells within the subependyma of the forebrain lateral ventricles, just as do other infraprimate mammals. However, whereas neural progenitors of the adult rodent and monkey ventricular wall generate neuronal progeny that migrate rostrally through the olfactory stream to the olfactory bulb, this pathway appears to be either absent or at least quantitatively vestigial in humans (Sanai et al 2004). Of likely greater importance in humans, the adult hippocampus harbours neuronal progenitor cells within the subgranular zone of the dentate gyrus, which may add neurons to the adult hippocampus in an ongoing process of granule cell replacement. Although the regulation and importance of hippocampal neurogenesis remain unclear in humans, in rodents it seems to be causally related to learning and memory acquisition. In addition, hippocampal neurogenesis may be involved in broader aspects of limbic function as well, with consequent implications with regards to human emotional regulation and affective disorders. Of even greater quantitative significance than dentate progenitors, there is an abundant population of parenchymal progenitor cells in the human forebrain; these cells are dispersed throughout the subcortical white matter, within which they appear to serve principally as glial progenitor cells (Nunes et al 2003). Just as in experimental animals, the glial progenitor cells of the adult human brain include a subpopulation of multipotential parenchymal progenitors, which though nominally glial may generate neurons as well as astrocytes and oligodendrocytes (Palmer et al 1999, Nunes et al 2003, Kondo & Raff 2000). As a whole, the glial progenitor is by far the most abundant precursor cell phenotype in the adult human brain, remarkably so, comprising over 4% of the cell population of the adult white matter.

Adult progenitors may be categorized as distinct pools of transit-amplifying cells

Over the past several years, several unifying concepts have evolved that describe the phenomenology, if not the common biology, of the different progenitor and stem cell phenotypes of adult organs. The concept of transit amplifying progenitors arising from less common pools of organ-specific stem cells has become an attractive conceptual framework for describing cell genesis and turnover in a variety of solid tissues. As initially defined in the skin and gastrointestinal (GI) mucosae, transit-amplifying cells comprise the phenotypically biased, still-mitotic progeny of uncommitted stem cells (Watt 2001, Niemann & Watt 2002, Loeffler & Potten 1997, Potten & Loeffler 1990). As stem cell progeny depart these localized regions of stem cell expansion, their daughters may commit to more restricted lineages, phenotypically delimited but still mitotic, which comprise the transit-amplifying pools. Although these cells proliferate so as to expand discrete lineages, they do not exhibit unlimited multilineage expansion, as distinct from their parental stem cells.

Within this conceptual framework, the neuronal and glial progenitor cells of the adult brain may be considered distinct transit-amplifying derivatives of a common ventricular zone stem cell (Doetsch et al 2002, Goldman 2003) (Fig. 1). The neuronal progenitor cell of the forebrain subependyma was first proposed as a transit-amplifying cell type, on the basis of its neuronal bias during mitotic expansion, and its ability to replenish the stem cell pool under appropriate mitotic stimulation (Garcia-Verdugo et al 1998). But since the neuronally committed progenitor cells of the rostral migratory stream, and those of the subgranular zone of the dentate gyrus, both continue to divide while migrating (Menezes et al 1995), they too comprise transit-amplifying phenotypes (Doetsch et al 2002). Indeed, even the glial progenitor cell of the adult white matter may now be considered a type of transit-amplifying cell, able to divide and yield variably restricted daughters, still mitotic but possessed of neither unbiased multipotentiality nor self-renewal capacity (Nunes et al 2003).

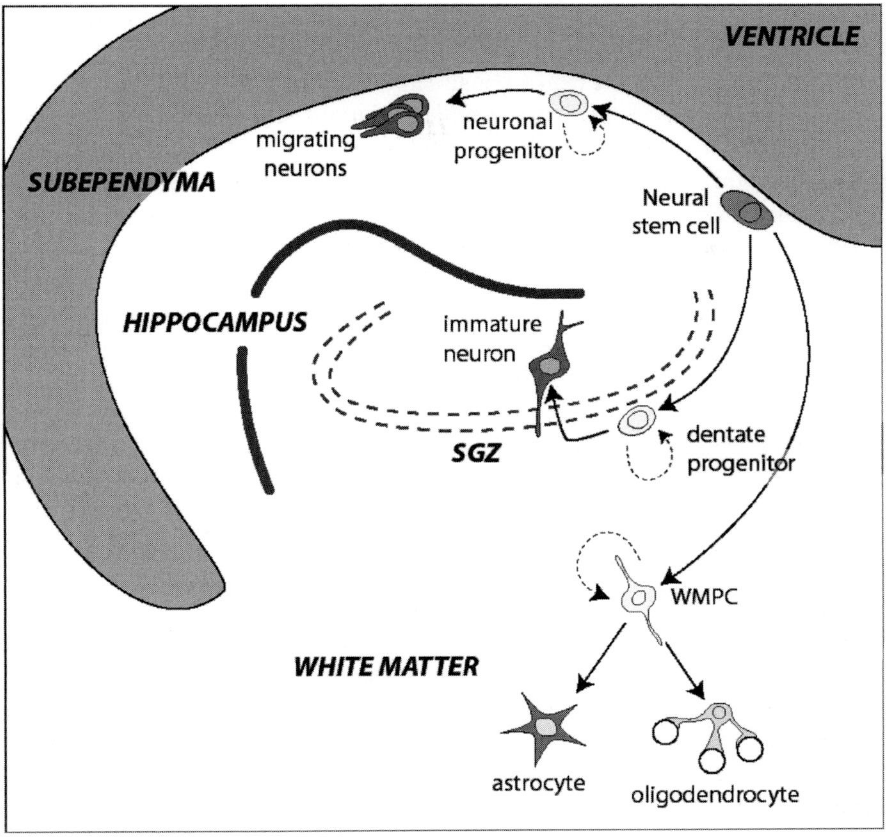

FIG. 1. Progenitor cells of the adult human brain. This schematic illustrates the identified categories of progenitors in the adult human brain, and their known interrelationships. All derive from ventricular zone neural stem cells, that generate at least three populations of potentially neurogenic transit-amplifying progenitors (neuronal progenitors, dentate progenitors and WMPCs) of both neuronal and glial lineages. These include neuronal progenitor cells of the ventricular subependyma, those of the subgranular zone of the dentate gyrus, and the white matter progenitor cells (WMPCs) of the subcortical parenchyma, which though nominally glial remain potentially neurogenic. Parenchymal progenitors may reside in the grey matter as well, although the relationships of parenchymal grey and white matter progenitors have yet to be established. Each transit-amplifying pool may give rise to progeny appropriate to their location, including neurons, oligodendrocytes and astrocytes. Adapted from Goldman (2003).

In adult humans, in whom the subependymal zone is but a discontiguous monolayer, ventricular zone neural stem cells appear to comprise a relatively scarce pool (Pincus et al 1998, Kirschenbaum et al 1994). Similarly, the neuronal progenitor pool of the olfactory subependyma appears to be similarly vestigial in humans (Sanai et al 2004), although the neuronal progenitors of the adult human hippocampus may be more abundant (Eriksson et al 1998, Roy et al 2000). As a result, in adult humans the major transit-amplifying pool would appear to reside within the parenchyma itself, in which large numbers of widely dispersed glial progenitors provide an abundant reservoir of cycling, multipotential progenitors, that though restricted to glial phenotype *in vivo*, are multipotential and neurogenic (Nunes et al 2003). As noted, these cells may comprise as many as 4% of all cells in the adult white matter, yielding remarkably high estimates for their absolute incidence; they comprise by far the most abundant progenitor phenotype of the adult brain (Roy et al 1999, 2004, Scolding et al 1999). Though the incidence of analogous parenchymal progenitor cells in the adult human grey matter has not yet been rigorously evaluated, these cells have been described as relatively abundant in the grey matter of adult rodents (Palmer et al 1999), and there is no reason to think that they are any less abundant in humans.

Parenchymal glial progenitors include multipotential neurogenic progenitors

The phenotypic definition of neural stem cells was made more difficult by recent observations that besides subependymal progenitors of the ventricular zone, the glial progenitors of the brain parenchyma might also have multilineage competence. Such multipotential precursor cells were first identified in the parenchyma of the adult rodent brain, as well as in its ventricular lining (Richards et al 1992, Palmer et al 1999). In the human, analogous pools were found within the cortex (Arsenijevic et al 2001) and subcortical white matter (Nunes et al 2003). The latter cells in particular, defined as white matter progenitor cells (WMPCs), were initially identified as oligodendrocyte progenitors on the basis of their expression of PDGFαR and the A2B5 epitope (Scolding et al 1998, 1999), their transcriptional activation of the CNP promoter (Roy et al 1999), and their predominant differentiation as oligodendrocytes, both *in vitro* and upon transplantation (Windrem et al 2002). Yet upon removal to low-density, serum-free culture, in which the cells are effectively removed from both autocrine and paracrine influences, WMPCs generate neurons as well as astrocytes and oligodendrocytes, and remain propagable for several months *in vitro*. Indeed, both P/CNP2:hGFP- and A2B5-sorted WMPCs can be separately propagated as neurospheres, expanding clusters of cells arising from single parental progenitors, whose clonally derived daughters include all major neural phenotypes (Nunes et al 2003) (Fig. 2).

Remarkably, the neurons derived from these white matter progenitors have proven functional, in that they develop normal and mature neuronal physiology, as assessed both by their calcium responses to depolarizing transmitters, and by whole-cell patch clamp analysis (Nunes et al 2003). Thus, the adult human white matter contains stem or progenitor cells that are able to divide, self-renew, and give rise to both neurons and glia within clonally expanding populations. Though biased to produce oligodendrocytes *in vivo*, these cells include multipotent cells that can be directed to neuronal and astrocytic lineage.

Glial progenitors of the subcortical white matter are transit-amplifying cells

In vivo, white matter progenitor cells appear as small, highly ramified cells with thin processes that lack endothelial endfeet or even contact. Most express nestin and the NG2 chondroitin sulfate proteoglycan, markers of immature neural progenitors, and S100β, a glial marker expressed by immature cells of both the astrocytic and oligodendrocytic lineages. Most express neither GFAP nor aquaporin, nor do they transcriptionally activate the GFAP promoter, and hence cannot be readily characterized as astroglial (Goldman 2003). Together, these data suggest that the glial progenitor of the adult human white matter, nominally an oligodendrocyte progenitor cell, is in fact a multipotential neural progenitor cell, restricted to generating glia by virtue of the adult parenchymal environment, and not because of any autonomous lineage commitment. Yet that is not to say that the WMPC is a stem cell. To the contrary, although parenchymal progenitor cells seem to be fundamentally multipotential, they are subject to replicative senescence, and do not express measurable telomerase (Nunes et al 2003). As a result, these cells typically cease expansion after 3–4 months *in vitro*, spanning no more than 18 population doublings. Thus, given their lack of telomerase, self-limited expansion capacity, and glial bias despite multilineage competence, adult white matter progenitor cells cannot be considered stem cells. Rather, they may best be considered transit-amplifying progenitors of both astrocytes and oligodendrocytes (Goldman 2003).

Just as the transit-amplifying neuronal precursor of the adult rat ventricular zone may revert to a multipotential state in the presence of epidermal growth factor (EGF) (Doetsch et al 2002), the dividing glial progenitor of the human white matter is similarly able to revert to a multilineal neurogenic precursor (Nunes et al 2003). Although this is especially manifest when the cells are expanded *in vitro*, adult glial progenitor cells may generate neurons as soon as they are removed from the local tissue environment, whether extracted on the basis of CNP promoter activation, A2B5 immunoselection (Nunes et al 2003), or NG2 expression (Belachew et al 2003). Accordingly, when freshly sorted parenchymal progenitors were introduced via transuterine xenograft into the fetal rat brain, all neural phenotypes were found to arise in a developmentally modulated and context-dependent manner (Nunes et al 2003) (Fig. 3). In general

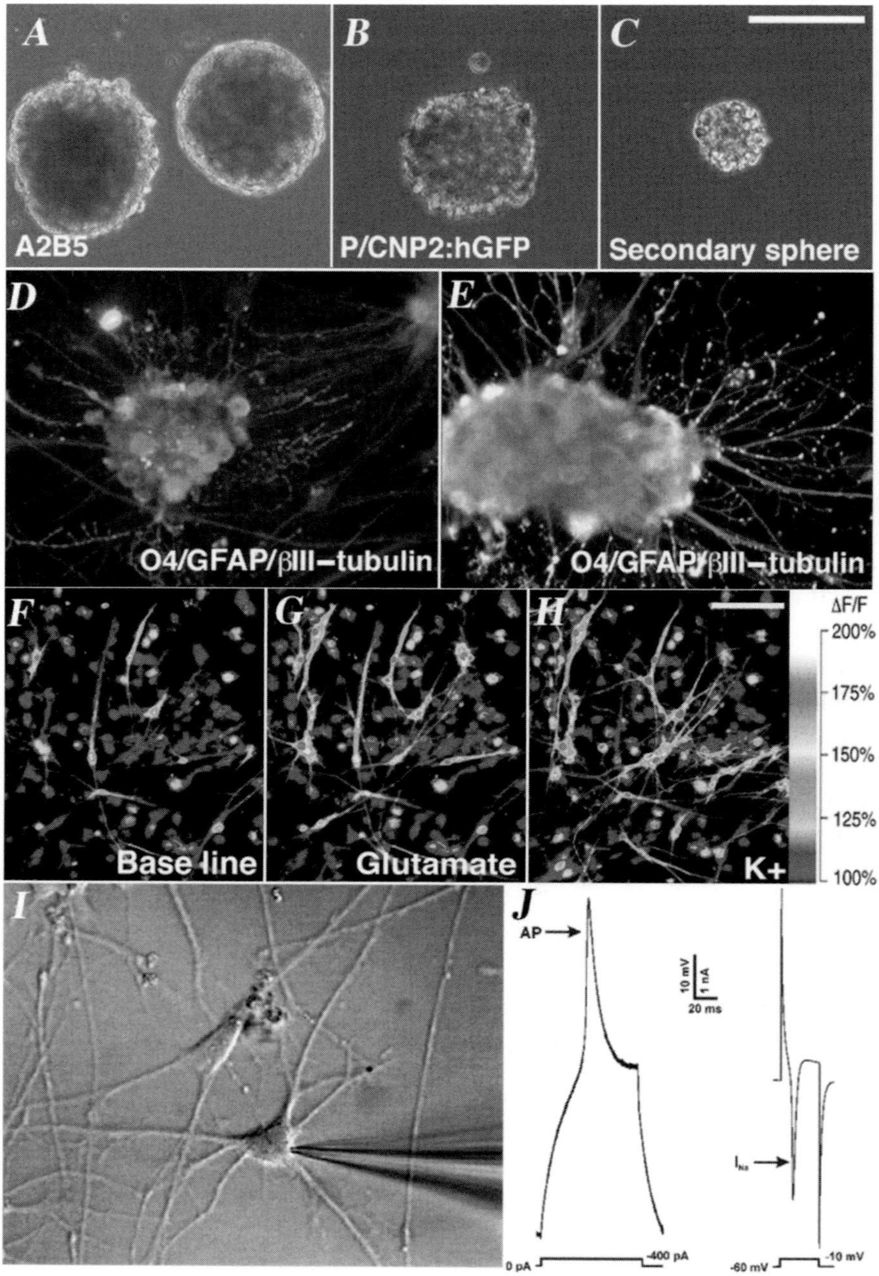
A
B
C
A2B5
P/CNP2:hGFP
Secondary sphere
D
E
O4/GFAP/βIII–tubulin
O4/GFAP/βIII–tubulin
F
G
H
ΔF/F
200%
175%
150%
125%
100%
Base line
Glutamate
K+
I
J
AP
10 mV
1 nA
20 ms
I_{Na}
0 pA
-400 pA
-60 mV
-10 mV

terms then, removal of glial progenitor cells from all environmental influences may prove both necessary and sufficient to ensure the appearance of multipotential progenitors in their resultant isolates. As such, one may consider the parenchymal progenitor pool of the adult brain as including both multipotential and glial-restricted members, the relative proportions of which are dynamically regulated by the local tissue environment. These observations suggest a hitherto unappreciated degree of cellular plasticity in the adult brain. A salient implication of this work is that with more understanding of the necessary and sufficient conditions for establishing neurogenic niches, we might expect to be able to modulate the parenchymal environment of the adult brain, to encourage the production of new neurons from resident parenchymal progenitor cells. Achieving and then directing this capability to therapeutic endpoints may prove an effort of great value. Indeed, just as our increasing understanding of the regulatory control of ventricular zone neural stem and progenitor cells has permitted the *in vivo* induction of striatal neurogenesis from resident progenitors (Benraiss et al 2001, Chmielnicki & Goldman 2002, Pencea et al 2001, Kuhn et al 1997, Aberg et al 2000, Louissaint et al 2002), we may anticipate the day when induced neurogenesis from resident progenitors in deep parenchymal structures might be similarly feasible.

Transplanted WMPCs disperse and engraft as oligodendrocytes in adult recipients

To assess the functional utility of adult human WMPCS, we have grafted them into a variety of models of both development and disease. We first chose to implant the cells into lysolecithin-demyelinated foci in the adult rat brain, so as to assess the myelinogenic competence of these nominally oligodendrocytic progenitors

---

FIG. 2. Adult human WMPCs generated neurons as well as glia. (A) Neurospheres arising from A2B5-sorted cells, 2 weeks post-sort. (B) First passage spheres raised from P/CNP2:GFP sorted cells, 2 weeks. (C) Second passage sphere derived from A2B5-sorted sample, 3 weeks. (D) Plated onto substrate, spheres differentiated as βIII-tubulin$^+$ neurons, GFAP$^+$ astrocytes and O4$^+$ oligodendrocytes (colours not visible in greyscale). (E) Neurons, astrocytes and oligodendrocytes arose from P/CNP2:GFP-sorted WMPCs (colours not discernable in greyscale). A–E, 100 μm. (F–H) WMPC-derived neurons had neuronal Ca^{2+} responses to depolarization (colour bar not represented in greyscale). (D) WMPC-derived cells loaded with the calcium indicator dye fluo-3, 10 days after plating first passage spheres derived from A2B5-sorted white matter (35 DIV); (E) after exposure to 100 μM glutamate; (F) depolarization by 60 mM KCl. Neurons displayed rapid, reversible, elevations in calcium in response to K^+. (F–H) 80 μm. (I) 14 days after plating a WMPC-derived sphere, neurons were subjected to whole cell patch clamp and their responses to current injection were noted. (J) (*left*), action potentials were noted at I_{Na} > 800 pA. (J) (*right*), fast negative deflections to current injection, typical of voltage-gated Na^+ currents of mature neurons. Adapted from Nunes et al (2003).

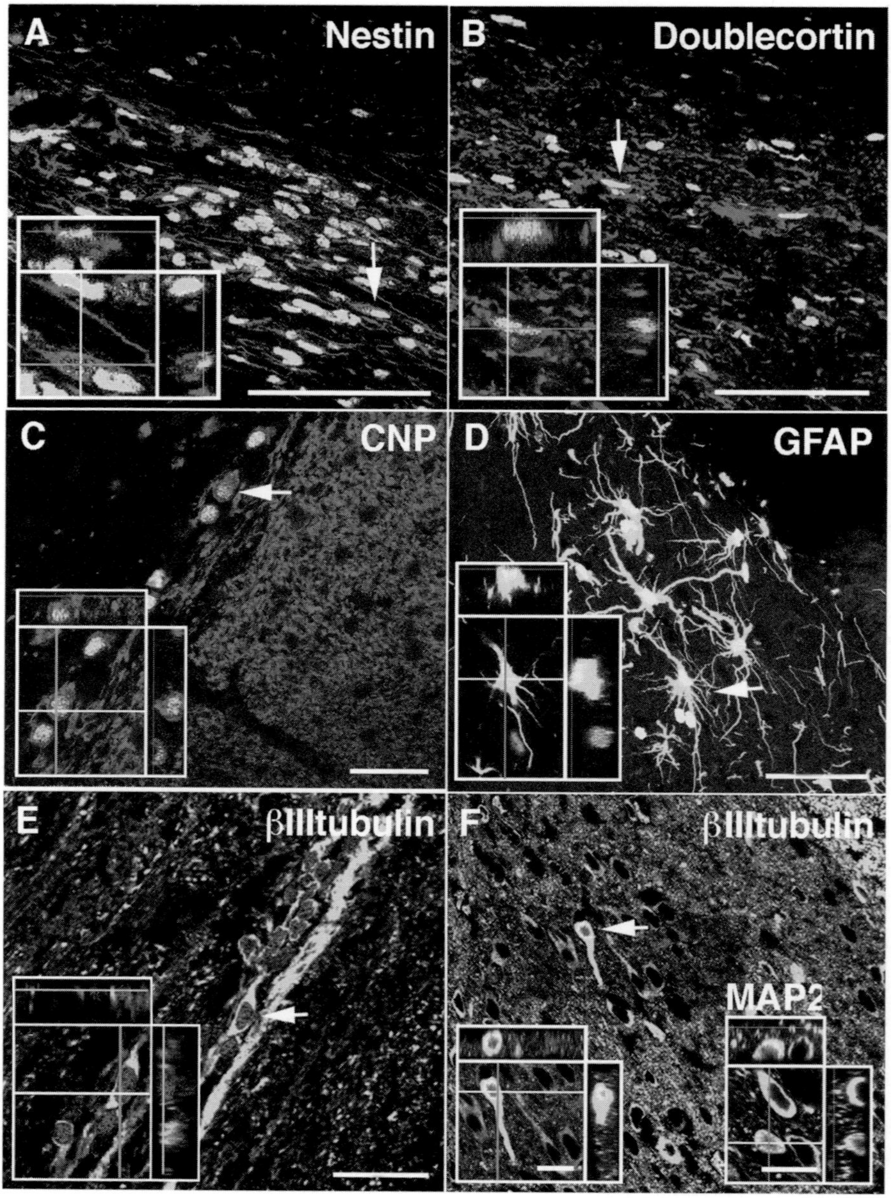
A
Nestin
B
Doublecortin
C
CNP
D
GFAP
E
βIIItubulin
F
βIIItubulin
MAP2

(Windrem et al 2002, 2003). In this set of experiments, 1×10^5 A2B5-sorted adult human WMPCs were transplanted into demyelinated lesions of the corpus callosa, 3 days after inducing these lesions by injecting the myelinotoxin lysolecithin. The implanted cells migrated throughout the extent of the demyelinated plaque, but not beyond its borders. As such, their migration appeared to be delimited by normal myelin, but unimpeded and rapid in the demyelinated regions. Most engrafted donor cells expressed CNP protein within two weeks of implant, indicating their oligodendrocytic maturation. By 3 weeks, many developed myelin basic protein (MBP) expression, suggesting progenitor-derived myelinogenesis within the lesion (Windrem et al 2003). With cyclosporine immunosuppression, these cells could survive at least 3 months in lysolecithin-demyelinated recipients. Thus, once isolated and extracted from the human white matter, adult WMPCs proved competent to mature and myelinate acutely demyelinated foci.

Adult WMPCs myelinate the congenitally dysmyelinated Shiverer brain

We next assessed the ability of human oligodendrocyte progenitors to engraft in a genetic model of congenital dysmyelination, the Shiverer mouse, which lacks expression of the gene encoding myelin basic protein (MBP). We found that A2B5-isolated WMPCs, derived from the adult white matter as well as from the late fetal subventricular zone, were both able to substantially myelinate the congenitally dysmyelinated shiverer forebrain: when these cells were transplanted into the presumptive corpus callosa of neonatal Shiverer mice, the implants yielded substantial remyelination throughout the subcortical white matter of the recipient shiverer mice (Windrem et al 2004). Indeed, a single perinatal injection of 100 000 cells into the corpus callosum dispersed by 4 weeks throughout the entire rostrocaudal and mediolateral extent of the callosum, corona radiata and proximal internal capsules. By 12 weeks, the donor-derived oligodendrocytes

FIG. 3. WMPCs engrafted into fetal rats generated neurons and glia in a site-specific manner. Sections from a rat brain were implanted at E17 with A2B5-sorted WMPCs, and sacrificed a month after birth. These cells were maintained in culture for 10 days prior to implant. A–B show nestin$^+$ (*light grey*) progenitors and doublecortin$^+$ (*dark grey*) migrants, respectively, each co-expressing human nuclear antigen (hNA, *light grey*) in the hippocampal alvius. C shows CNP$^+$ oligodendrocytes (*dark grey*), that were found exclusively in the corpus callosum. D, a low-power image of GFAP$^+$ (*light grey, stained with anti-human GFAP*) astrocytes along the ventricular wall. E, βIII-tubulin$^+$ (*light grey*)/hNA$^+$ (*dark grey*) neurons migrating in a chain in the hippocampal alvius. F, βIII-tubulin$^+$ and MAP2$^+$ (*inset* in F) neurons in the striatum, adjacent to the RMS (antigens in *light grey*; hNA in *dark grey*; *pale grey rings* double-stained human nuclei). Scale: A–E, 40 μm; F, 20 μm. Taken from Nunes et al (2003).

had generated dense sheaths of compact myelin, which surrounded up to 10% of host callosal and capsular axons (Fig. 4). Untreated Shiverer mice are characterized by a complete lack of MBP expression, so that the widespread myelination observed in our treated animals was necessarily donor-derived. In addition, the WMPC-derived oligodendrocytes exhibited a normal paranodal distribution of the nodal marker protein Caspr along donor-ensheathed axons. Engrafted adult human OPCs thereby developed not only normal myelin architecture, but structurally appropriate nodes of Ranvier, suggesting their functional integrity

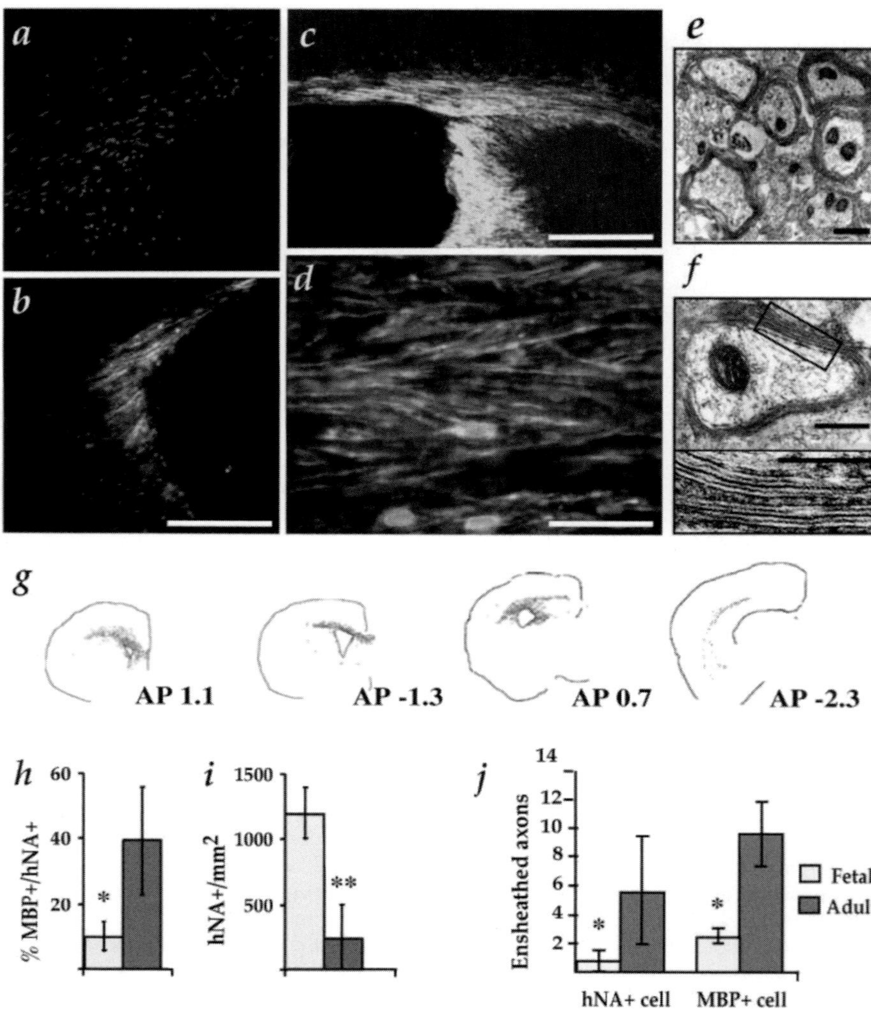

(M. S. Windrem, S. Goldman, unpublished results). Thus, adult WMPCs can efficiently differentiate as myelinogenic oligodendrocytes after removal and transplantation to the subcortical white matter environment of a heterologous host (Roy et al 2004). This observation suggests the potential of human oligodendrocyte progenitor cell grafts as a therapeutic modality for both congenital and acquired demyelinating diseases.

Identification of transcripts differentially expressed by adult human WMPCs

Neural progenitors of the adult brain may serve not only as cellular substrates for transplantation, but also as targets for endogenous activation and pharmacological induction. To define a set of potential targets by which to modify the activity and fate of these cells *in vivo*, we chose a genomics-based strategy to define the receptors and signalling systems utilized by adult progenitor cells in their normal resting state. To that end, we have used oligonucleotide microarrays to analyse the transcriptional profile of A2B5$^+$ human WMPCs, separately sorted from four surgically resected adult temporal lobes. From each of the four temporal lobe

FIG. 4. Engrafted human WMPCs myelinated an extensive region of the forebrain (a) Adult-derived human WMPCs (hNA, *dark grey*) achieved dense MBP expression (*light grey*) by 4 weeks after xenograft. (b) In contrast, fetal WMPCs expressed no detectable MBP-IR at 4 weeks, and little until 12 weeks. Fetal and adult WMPCs were found to differ substantially in their speed and efficiency of myelinogenesis (discussed in Windrem et al 2004). (c–d) Low and high magnification coronal images of the callosal-fimbrial junction of a shiverer homozygote, showing dense myelination by 12 weeks after perinatal engraftment with adult-derived human WMPCs. When assessed individually, almost half of the donor cells in this recipient white matter were found to express MBP. (e–f) Adult WMPCs developed mature myelin ultrastructure and major dense lines within 5 weeks of perinatal injection. **e** and **f** show examples of myelin formed by implanted WMPCs in the shiverer callosum, in an animal sacrificed 5 weeks after birth, who had been given an intraventricular injection of adult WMPCs on day 1 after birth. The myelin structure seen here was analogous to that of fetal OPCs assessed 12 weeks after cell injection. Again, the mice injected with fetal OPCs exhibited no such evidence of myelination at this early timepoint. **g** shows the distribution of adult WMPC donor cells 4 weeks after perinatal implantation into a single site within the corpus callosum; the distribution of human NA$^+$ cells in this image reflects the widespread dispersal of the cells through the presumptive white matter tracts. **h**, A substantially higher proportion of implanted adult OPCs developed MBP expression than did fetal OPCs, when both were assessed at 12 weeks. **i**, Fetal donor cells nonetheless engrafted more efficiently and in higher numbers than did identically-implanted adult OPCs. *indicates $P<0.05$; **$P<0.005$, each of Student's *t*-test (two-tailed). **j**, This graph plots the numbers of ensheathed axons per donor cell achieved by both fetal and adult-derived OPCs. Ensheathment was defined by confocal-imaged MBP$^+$ enwrapment of NF$^+$ axons, and was measured as a function both of total donor cell number (*left* panel), and of MBP$^+$ donor-derived oligodendrocytes (*right*). From Windrem et al 2004. Scale: A,B, 100 μm; C, 1 mm; D, 30 μm.

resections, we obtained between 5×10^5 and 1×10^6 A2B5$^+$ cells that comprised 3–4% of all viably dissociated white matter cells. Using U95A Affymetrix chips, we then obtained the individual gene expression profiles of each these A2B5$^+$-sorted samples, and normalized each against that of the unsorted white matter dissociate from which it was derived. By this means, we were able to identify a set of WMPC-enriched transcripts, otherwise under-represented in the central white matter. The resulting list of approximately 250 probe sets — comprising $<5\%$ of the more than 8500 genes screened on the U95A — was edited by removing those that were either ambiguously annotated as mapping to multiple genes, or that were novel, by virtue of not yet having been annotated to NCBI LocusLink identifiers (*www.ncbi.nih.gov/LocusLink*). The remaining probe sets were annotated to 210 distinct genes that were selectively overexpressed by the sorted WMPCs. A total of 51 genes were also identified as selectively underexpressed by the WMPCs, relative to their parental white matter (Sim et al 2004).

To analyse the microarray data, we first examined the expression of several marker genes previously described as differentially expressed by glial progenitor cells (Fig. 5). The antibody used to isolate adult WMPCs, monoclonal A2B5 (Raff et al 1983, 1984, Roy et al 1999, Eisenbarth et al 1979), primarily recognizes the gangliosides G_Q, G_{T3} and their O-acetylated derivative (Farrer & Quarles 1999). We found that the expression of GD3 synthase (SIAT8A), the enzyme that catalyses the transfer of sialic acid from CMP-sialic acid to GM3 by which GD3 and GT3 are generated, was significantly enriched in the A2B5-isolated WMPC pool. This observation was confirmed with real-time RT-PCR analysis (qPCR) of GD3 synthase mRNA levels following normalization to 18S ribosomal RNA. The isolated WMPCs also expressed high levels of PDGFαR and NG2 (CSPG4), two canonical markers of oligodendrocyte progenitors *in vivo*, which we also confirmed by qPCR. This expression pattern was consistent with their nominal oligodendrocyte progenitor phenotype. As would have been expected, more mature oligodendrocytic transcripts, including CNP and the myelin protein genes, myelin basic protein (MBP) and proteolipid protein (PLP1), were under-expressed by WMPCs, relative to their parental white matter, as were markers for other cells resident in the white matter, including astrocytes, microglia and endothelial cells. Surprisingly though, the sorted adult WMPCs selectively expressed high levels of transcripts typical of undifferentiated neural progenitor cells, such as HES1 and musashi, as well as early pro-neural genes such as MASH1. The differential expression of these transcripts, as well as that of a number of members of the Notch pathway, suggested that WMPCs were indeed pre-oligodendrocytic while resident *in vivo*. Furthermore, they also differentially expressed several cognate pairs of ligands and receptors, suggesting prominent autocrine regulation when quiescent. Together, these data suggest that parenchymal progenitor cells of the adult human brain comprise a fundamentally

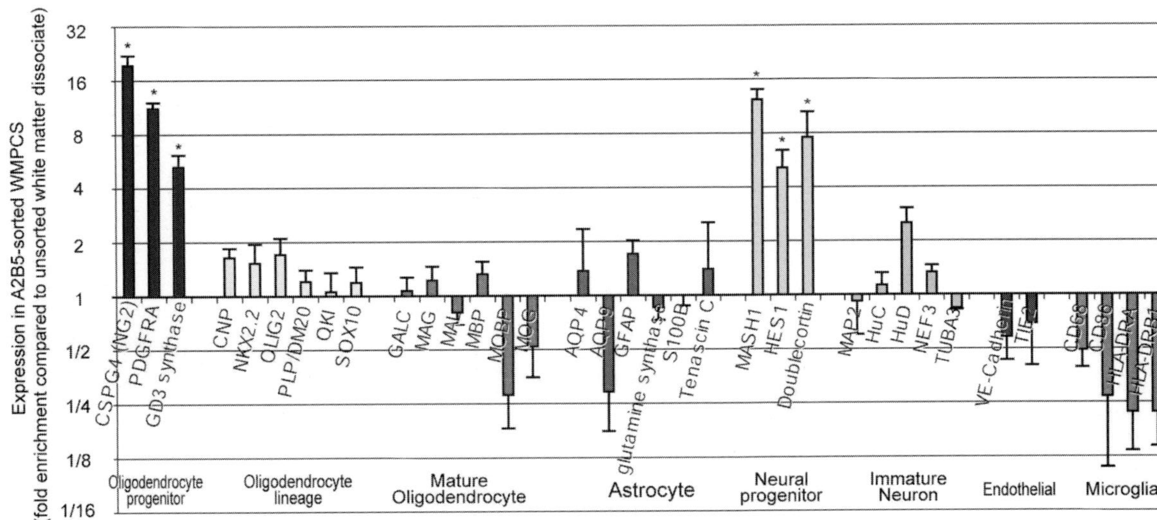

FIG. 5. Adult human WMPCs express neural progenitor marker genes, and are transcriptionally distinct from the local white matter environment. Microarray analysis of sorted adult WMPCs revealed that, relative to their parental white matter, these cells selectively expressed transcripts typical of oligodendrocyte progenitors, such as NG2 and PDGFαR. in addition, they were characterized by their co-expression of transcripts typical of both undifferentiated neural progenitors, such as HES1, and early pro-neural genes such as MASH1 and doublecortin.

primitive phenotype, undifferentiated *in situ* yet potentially subject to environmental override of an autocrine homeostasis. As such, induced neurogenesis and oligoneogenesis from these cells may prove both an attractive and potent therapeutic strategy for a variety of diseases of acquired neurodegeneration and demyelination. Indeed, given the abundance and widespread distribution of these primitive neural progenitors, and their lineal relationship to other progenitor phenotypes of the adult brain, we can be assured that achieving the regulatory control of these cells will be an important goal of neurological therapeutics for years to come.

Acknowledgments

We would like to thank our collaborators in these studies, in particular Drs Neeta Roy, Martha Windrem, H. Michael Keyoung, Marta Nunes, Abdellatif Benraiss, Su Wang and Maiken Nedergaard. Our work has been supported by the NIH, the Mathers Charitable Foundation, the National Multiple Sclerosis Society, the Human Frontiers Scientific Program, the Michael J. Fox Foundation, Christopher Reeve Paralysis Foundation, Project ALS, the New York Spinal Cord Injury Program, and by both Aventis Pharmaceuticals and Merck, Inc.

References

Aberg M, Aberg D, Hedbacker H, Oscarsson J, Eriksson P 2000 Peripheral infusion of IGF-1 selectively induces neurogenesis in the adult rat hippocampus. J Neurosci 20:2896–2903

Alvarez-Buylla A, Garcia-Verdugo JM 2002 Neurogenesis in adult subventricular zone. J Neurosci 22:629–634

Arsenijevic Y, Villemure J, Brunet J et al 2001 Isolation of multipotent neural precursors residing in the cortex of the adult human brain. Exp Neurol 170:48–62

Belachew S, Chittajallu R, Aguirre A et al 2003 Postnatal NG2 proteoglycan-expressing progenitor cells are intrinsically multipotent and generate functional neurons. J Cell Biol 161:169–186

Benraiss A, Chmielnicki E, Lerner K, Roh D, Goldman SA 2001 Adenoviral brain-derived neurotrophic factor induces both neostriatal and olfactory neuronal recruitment from endogenous progenitor cells in the adult forebrain. J Neurosci 21:6718–6731

Chmielnicki E, Goldman SA 2002 Induced neurogenesis by endogenous progenitor cells in the adult mammalian brain. Prog Brain Res 138:451–464

Doetsch F, Alvarez-Buylla A 1996 Network of tangential pathways for neuronal migration in adult mammalian brain. Proc Natl Acad Sci USA 93:14895–14900

Doetsch F, Garcia-Verdugo JM, Alvarez-Buylla A 1997 Cellular composition and three-dimensional organization of the subventricular germinal zone in the adult mammalian brain. J Neurosci 17:5046–5061

Doetsch F, Petreanu L, Caille I, Garcia-Verdugo J, Alvarez-Buylla A 2002 EGF converts transit-amplifying neurogenic precursors in the adult brain into multipotent stem cells. Neuron 36:1021–1034

Eisenbarth GS, Walsh FS, Nirenberg M 1979 Monoclonal antibody to a plasma membrane antigen of neurons. Proc Natl Acad Sci USA 76:4913–4917

Eriksson PS, Perfilieva E, Bjork-Eriksson T et al 1998 Neurogenesis in the adult human hippocampus. Nat Med 4:1313–1317

Farrer RG, Quarles RH 1999 GT3 and its O-acetylated derivative are the principal A2B5-reactive gangliosides in cultured O2A lineage cells and are down-regulated along with O-acetyl GD3 during differentiation to oligodendrocytes. J Neurosci Res 57:371–380
Gage F 2000 Mammalian neural stem cells. Science 287:1433–1438
Garcia-Verdugo J, Doetsch F, Wichterle H, Alvarez-Buylla A 1998 Architecture and cell types of the adult subventricular zone: in search of the stem cells. J Neurobiol 36:234–248
Goldman S 1998 Adult neurogenesis: from canaries to the clinic. J Neurobiol 36:267–286
Goldman S 2003 Glia as neural progenitor cells. Trends Neurosci 26:590–596
Kirschenbaum B, Goldman SA 1995 Brain-derived neurotrophic factor promotes the survival of neurons arising from the adult rat forebrain subependymal zone. Proc Natl Acad Sci USA 92:210–214
Kirschenbaum B, Nedergaard M, Preuss A et al 1994 In vitro neuronal production and differentiation by precursor cells derived from the adult human forebrain. Cerebral Cortex 4:576–589
Kondo T, Raff M 2000 Oligodendrocyte precursor cells reprogrammed to become multipotential CNS stem cells. Science 289:1754–1757
Kornack D, Rakic P 1999 Continuation of neurogenesis in the hippocampus of the adult macaque monkey. Proc Natl Acad Sci USA 96:5768–5773
Kuhn HG, Winkler J, Kempermann G, Thal LJ, Gage FH 1997 Epidermal growth factor and fibroblast growth factor-2 have different effects on neural progenitors in the adult rat brain. J Neurosci 17:5820–5829
Louissaint A Jr, Rao S, Leventhal C, Goldman SA 2002 Coordinated interaction of neurogenesis and angiogenesis in the adult songbird brain. Neuron 34:945–960
Luskin MB 1993 Restricted proliferation and migration of postnatally generated neurons derived from the forebrain subventricular zone. Neuron 11:173–189
Menezes JR, Smith CM, Nelson KC, Luskin MB 1995 The division of neuronal progenitor cells during migration in the neonatal mammalian forebrain. Mol Cell Neurosci 6:496–508
Morshead CM, Reynolds BA, Craig CG et al 1994 Neural stem cells in the adult mammalian forebrain: a relatively quiescent subpopulation of subependymal cells. Neuron 13:1071–1082
Niemann C, Watt FM 2002 Designer skin: lineage commitment in postnatal epidermis. Trends Cell Biol 12:185–192
Nunes MC, Roy NS, Keyoung HM et al 2003 Identification and isolation of multipotential neural progenitor cells from the subcortical white matter of the adult human brain. Nat Med 9:439–447
Palmer TD, Takahashi J, Gage FH 1997 The adult rat hippocampus contains primordial neural stem cells. Mol Cell Neurosci 8:389–404
Palmer TD, Markakis EA, Willhoite AR, Safar F, Gage FH 1999 Fibroblast growth factor-2 activates a latent neurogenic program in neural stem cells from diverse regions of the adult CNS. J Neurosci 19:8487–8497
Pencea V, Bingaman KD, Wiegand SJ, Luskin MB 2001 Infusion of brain-derived neurotrophic factor into the lateral ventricle of the adult rat leads to new neurons in the parenchyma of the striatum, septum, thalamus, and hypothalamus. J Neurosci 21:6706–6717
Pincus DW, Keyoung HM, Harrison-Restelli C et al 1998 Fibroblast growth factor-2/brain-derived neurotrophic factor-associated maturation of new neurons generated from adult human subependymal cells. Ann Neurol 43:576–585
Potten CS, Loeffler M 1990 Stem cells: attributes, cycles, spirals, pitfalls and uncertainties. Lessons for and from the crypt. Development 110:1001–1020
Raff MC, Miller RH, Noble M 1983 Glial cell lineages in the rat optic nerve. Cold Spring Harb Symp Quant Biol 48:569–572
Raff MC, Williams BP, Miller RH 1984 The in vitro differentiation of a bipotential glial progenitor cell. EMBO J 3:1857–1864

Richards LJ, Kilpatrick TJ, Bartlett PF 1992 De novo generation of neuronal cells from the adult mouse brain. Proc Natl Acad Sci USA 89:8591–8595
Roy NS, Wang S, Harrison-Restelli C et al 1999 Identification, isolation, and promoter-defined separation of mitotic oligodendrocyte progenitor cells from the adult human subcortical white matter. J Neurosci 19:9986–9995
Roy NS, Wang S, Jiang L et al 2000 In vitro neurogenesis by progenitor cells isolated from the adult human hippocampus. Nat Med 6:271–277
Sanai N, Tramontin A, Quinones-Hinojosa A et al 2004 Unique astrocyte ribbon in adult human brain contains neural stem, cells but lacks chain migration. Nature 427:740–743
Scolding N, Franklin R, Stevens S et al 1998 Oligodendrocyte progenitors are present in the normal adult human CNS and in the lesions of multiple sclerosis. Brain 121:2221–2228
Scolding NJ, Rayner PJ, Compston DA 1999 Identification of A2B5-positive putative oligodendrocyte progenitor cells and A2B5-positive astrocytes in adult human white matter. Neuroscience 89:1–4
Watt FM 2001 Stem cell fate and patterning in mammalian epidermis. Curr Opin Genet Dev 11:410–417
Weiss S, Dunne C, Hewson J et al 1996 Multipotent CNS stem cells are present in the adult mammalian spinal cord and ventricular neuroaxis. J Neurosci 16:7599–7609
Windrem MS, Roy NS, Wang J et al 2002 Progenitor cells derived from the adult human subcortical white matter disperse and differentiate as oligodendrocytes within demyelinated lesions of the rat brain. J Neurosci Res 69:966–975
Windrem MS, Nunes MC, Rashbaum WK et al 2004 Fetal and adult human oligodendrocyte progenitor cell isolates myelinate the congenitally dysmyelinated brain. Nat Med 10:93–97

DISCUSSION

Björklund: I'd be interested to know more about the white matter progenitor cells (WMPCs). How essential is the cell sorting in order to be able to isolate them. Can you isolate them directly from the primary cell material?

Goldman: Yes, that is how we do it. Typically, we take the white matter, dissociate it and either transfect with the CNP2 promoter-driven GFP, or use a surrogate marker which we have identified—the G_Q and GD3 gangliosides expressed by the cells are recognized by the A2B5 antibody. We can select WMPCs by using A2B5 followed by PSA-NCAM depletion to deplete the sample of any early neuronally committed cells. This combination of A2B5 positive selection followed by PSA-NCAM depletion gives essentially the same phenotype as the promoter sort. This is actually faster, because the promoter sort requires transfection and then a couple of days in culture for the green fluorescent protein (GFP) to be expressed. We grow the cells in that case in deficient media and in suspension, but there is a culture time of 36–48 h before sorting. By way of comparison, with A2B5 immunomagnetic selection we can go from a piece of tissue to a purified preparation of isolated progenitor cells in a few hours.

Björklund: Does this also involve fluorescence-activated cell sorting (FACS)?

Goldman: We can do it both ways. To make it really quick we use immunomagnetic sorting followed by high speed FACS depletion of the PSA-

NCAM pool. You can also do it as a two-colour FACS but the yield falls and it is slower to do.

Björklund: This means that you use an enrichment step before you grow them up. What happens if you don't do that? Can you get these kinds of spheres using your growth factor combination with platelet-derived growth factor (PDGF) and fibroblast growth factor (FGF) on the primary cell preparation?

Goldman: That is a good question. The basic answer is no. Spheres are generated, but we can't passage them and they give rise to astrocytes.

Björklund: Is this PDGF dependent? And how do they relate to standard neurospheres? Once you have sorted these could you make them proliferate in the standard EGF/FGF/leukaemia inhibitory factor (LIF) mixture?

Goldman: We have *not* found LIF helpful here. We have made the direct comparison to FGF alone, and the clonogenic incidence remains as high, but the incidence of neurogenic spheres and oligogenic spheres is much lower. The proportion of the clones that are tripotential is much lower in FGF alone, compared to PDGF and NT3 as well. We don't know whether the addition of PDGF is helping to select for a different population within the overall white matter cell pool, or whether it is actually broadening the differentiation potential of the cell.

Björklund: What would happen if you made the neurosphere expansion first and identified your sub-population later? Would this sub-population of progenitors be contained within the standard neurospheres?

Goldman: I predict that they would be. We are looking at this not from the standpoint of the efficiency being any better, but because we want to be able to remove the progenitors from any *ex vivo* autocrine or paracrine factors from the outset. In general, I suspect that this is why we have had better success with our strategy, in that the progenitor cells are being excluded from astrocyte-derived factors and so on, as soon as they are removed from the tissue.

Jaenisch: What I understood is that you used transfection with a plasmid which drives GFP from a tubulin promoter to introduce a genetic label into the cells. Then whatever lights up you define as a neural stem cell. Is that correct?

Goldman: No, the sorting is done on the basis of GFP expression, where GFP has been placed under the control of promoters or enhancers that direct gene expression to neural stem or progenitor cells. For this we use regulatory sequences that can specify gene expression to the cell type of interest. Neuronal progenitor cells drive expression of Tα1 tubulin, a finding made years ago by Freda Miller's group. We verified that the Tα1 tubulin promoter can direct gene expression to neuronal progenitor cells. We also found that an intronic enhancer of the nestin gene, that was first described by Ron McKay's group, can similarly direct gene expression to neural stem cells, along with immature neurons and glia.

Jaenisch: Is it a transient transfection? There are two potential problems: first, you don't know where the transgenes integrate. Then you don't know the transfection efficiency.

Goldman: We don't want the transgene to integrate. The whole idea here is to use a transient transfection to recognize the cell type of interest, sort it, and then we're done with the genetic tag. The promoter–GFP construct is just a tool that we're using for the separation; we're not trying to permanently tag the cells. We have done this with a plasmid and a virus. With the plasmid it depends on the cell type and plasmid size. For the specific phenotype I am concentrating on here, the white matter progenitor, the transfection efficiency using a CNP2-driven GFP is just under 15%. We lose the other 85% that we would predict that would be in the population. We account for this in our calculations of the predicted incidence of each cell type. To get around this in later work, we took the same promoter–GFP selection vectors and put them into adenoviruses. Our protocol is to take the viruses and infect the cells in suspension as soon as the tissue has been dissociated. The viral infection efficiency is much higher, and runs 80–90% at doses where we still don't see any loss in viability from the viral infection.

Jaenisch: But you are defining your stem cell as being able to express a plasmid that got in. I am uncomfortable with this criterion. Transfecting cells with GFP is not a good way to mark them. How many cells are transfected and don't express the plasmid? Are the same number of copies present?

Goldman: You have raised a couple of points here that we've addressed by means of our controls. The transfection efficiency is defined on the basis of the proportion of all cells that express GFP at a defined time point after transfection, usually 2 days. To define the net transfection efficiency, we use a constitutive promoter-driven GFP, so that all successfully transduced cells should become GFP+. We use the CMV promoter for this purpose, since it drives strong gene expression in all neural cell types. The result is that we can define the transfection efficiency in a relatively straightforward fashion with constitutive promoter-driven GFPs. These essentially define our upper limit. Of course, we then don't know what the incidence of the target cell is within the original population. In order to determine this we then use a selection construct-driven GFP, and determine its incidence of expression in the cell population. Dividing the incidence of specific promoter-driven GFP+ cells, by that of constitutive promoter-driven GFP+ cells in matched cultures then gives you an estimate of the incidence of your target cells in the larger cell population. Let's use separating neural progenitor cells on the basis of nestin–GFP expression as an example. We start by doing two controls: the first is a positive control, using CMV-driven GFP to define the transfection efficiency. Let's say 15% of the cells in our culture express GFP after CMV-GFP transfection. That means that 15 out of 100 cells in the culture were successfully transduced by the plasmid, assuming 100% efficiency of CMV promoter-dependent transcription in these cells. We transfect a matched culture with nestin-GFP, and find that 1% of the cells are GFP+. What this means is that 1 cell out of 15 in that culture is driving nestin, and may be defined on that basis as a progenitor cell, at least in so far as the nestin enhancer used is specifically activated in neural progenitor cells. As a negative control, we also transfect the cells with a GFP expression plasmid placed under the control only of a weak basal promoter, and we use this to define the leakiness of the system, in other words the incidence of false positives. With this set of numbers, including the CMV-defined transfection efficiency, the nestin-GFP incidence, and the false positive incidence, we can predict the absolute incidence of our target cell type in the population. So with this approach, I think we are reasonably well covered.

I think that your point is more important with regards to copy number. We see a lot of spread in terms of fluorescence intensity when we FACS. This may indeed be an artefact of copy number, and there is nothing that we can do about this. What we do in practice is adjust the threshold at which we are sorting to a much higher level of stringency than the controls, to account for exactly that.

Let's go beyond the technology for a second, because this gets to the underlying rationale for using a promoter-based selection strategy from cell type identification. This is first and foremost an empiric strategy for assigning phenotype. We are defining a cell type on the basis of its transcriptional activation of a specific regulatory region, after we have already defined that region as specifying gene expression to a given cell type, based on both gene expression patterns, and on transgenic and somatic promoter reporter assays. But after having done this, the proof is in the pudding: we then verify the cells we have defined empirically, let's say as neural stem cells based on nestin–GFP expression, by analyzing how they behave and the cells they give rise to after their isolation. Ultimately, these assessments of phenotype are functional and physiological, and serve to validate the cell-type specificity of the promoter selection strategy.

Jaenisch: To me this is a circular argument. It is an unsatisfying lineage marker, because it is defined by transfection, which is rather artificial.

Goldman: No, don't confuse empiricism with circularity. This approach provides an empiric definition of cell type, based on transcriptional activity of transgenes encoding reporters driven by non-coding regulatory regions. These are simply being introduced into the cells by transfection, that's all. The plasmids may be synthetic, but there's nothing artificial about the logic.

Jaenisch: I could argue that what you see are the most easily transfectable cells with the highest copy number but these have no relevance to stem cells.

Goldman: So what? Sure, copy number is a potential source of artefact, because it introduces variability in the degree of fluorescence of positive cells, in which the promoter–GFP construct is active. But negatives, in other words undesired cell types, in which the introduced promoter–GFP is inactive, still fail to express the GFP, and hence are not recognized. So although variability in copy number can confuse the degree to which a given cell expresses from a given promoter, it does not significantly alter the rate of either false positives or false negatives in the selection steps, or the accuracy of the cell selection achieved. This is more of a technical issue in the FACS procedure than anything else.

I should mention here that our experience with this strategy is not limited to nestin, Tα1 or CNP2. We've set up over a dozen of these selection constructs for different cell types. Our goal is to profile the precursors recognized and isolated by each, to see what is different about their respective molecular signatures, in part so that we can use this information to predict surface antigenic phenotypes. We have also set up transgenics for most of these to see which phenotype is being recognized *in vivo* by each selection construct. For nestin, CNP2 and Tα1, the phenotypes recognized *in vivo* in their respective transgenics have been exactly those that we've sorted. But we've had other candidate promoters whose reporter expression patterns in transgenics have not turned out to be as specific as we'd have hoped. In those cases, we've not pursued those as selection vectors.

Minger: In the Shiverer neonates you get very extensive myelination with the white matter progenitor cells. Have you tried this in adult Shiverer mice?

Goldman: No. We did a couple of pilots where we looked at the extent of migration, which was not nearly as effective as in the perinatals.

Solter: Can one say that there probably are stem cells in the adult brain, which can be taken out and shown in culture to be stem cells? Is there a logic behind what Pasko Rakic suggested in a recent review that in adult brain although they are present, they don't actually do anything (Rakic 2004). If so, we can predict that if we inject stem cells into adult brain they are not going to do anything, just as the endogenous stem cells don't do anything? Is there logic behind this?

Goldman: The logic is excellent, since the issue is one of environmental restriction, rather than of the capabilities of the stem cells themselves. The restriction here seems to be imposed by the adult brain environment. The ventricular zone stem cell population in the adult normally does not appear to produce neurons that actually enter the brain parenchyma. Instead, when stem cell-derived neurons go to the olfactory bulb, they stay within the plane of the olfactory subependyma, which is a direct extension histologically and developmentally of the ventricular zone. In early development the telencephalic vesicle essentially zips up. So in the adult rodent, in which neurons are constantly generated and move to the olfactory bulb, these cells never actually leave the ventricular zone. It is only in birds that they leave the ventricular wall to go into the parenchyma. In contrast, in the normal adult mammal, there's no departure of newly generated neurons from the ventricular wall into the underlying parenchyma. There is either migration forward to the olfactory bulb, or back to the subgranular zone and then the dentate gyrus and hippocampus. To use stem cells therapeutically then, whether we're talking about endogenous progenitors or transplanted cells, we need to provide a permissive environment for migration and departure from the ventricular zone, while modulating the phenotypes generated at any given point within the ventricular wall. In fact, in adult humans there is an abundant progenitor population throughout the white and grey matter. But there is not a shred of evidence that in normal physiology or disease that these cells do anything but give rise to glia. In lower species neurons generated from these cells in the neurogenic regions subserve dentate gyrus function, hippocampal functions and olfactory bulb neurogenesis. In humans, the olfactory bulb is essentially vestigial. So I don't think that there is any strong argument for the fundamental biology of the human and the rat being any different. Neither appears to allow ventricular zone neuronal departure in adults, at least not in healthy and uninjured brain.

This discussion brings up a few points, for which I'd like to discuss some additional data. The issue here is how to mobilize endogenous pools and get them to leave the ventricular zone. We injected a virus overexpressing brain-derived neurotrophic factor (BDNF), which we have shown is a differentiation factor driving stem cells to neurons, into the ventricular wall. The virus maintains expression of around 2 ng/ml CSF for around two months. The aim of the experiment was to see what the effect of BDNF would be on cells being generated from the ventricular wall (in other words, the ventricular stem cell population that would normally give rise to neurons going to olfactory bulb or to glia). So we gave BrDU every day as a mitotic marker for three weeks after viral delivery, and then looked at what was happening in those cells generated in the ventricular zone in the presence or absence of BDNF overexpression. We essentially turned the ependyma into a secretory source for the neurotrophic factor that would not normally be seen by subependymal cells. We scored BrDU-expressing cells in these regions and saw a substantial increase in neurogenesis to the olfactory bulb—no surprise. But we were hoping to see heterotopic neurogenesis and migration elsewhere. We didn't see this to the cortex or septum, but we did in the striatum. This makes some sense because the greatest density of ventricular zone progenitors is along the striatal wall. Here this olfactory bulb population is now being hit by BDNF before it normally would be, and looking at the striatum we can identify neurons of the major sub-phenotypes of medium spine neurons. In the caudate and putamen, the medium spiny neurons are by far the most abundant subpopulation. This turns out to be closely related to the olfactory bulb neuronal population. These cells can be induced to leave the ventricular zone, to be mobilized, and then recruited into the adult striatum. If we quantify this, we find about 150 neurons per cubic millimetre when we add BDNF, which is very encouraging but may not be enough to anticipate a functional contribution. If we are to have any therapeutic effect we probably need to do better. From the array studies on assorted progenitors we have identified a number of BMPs that are being made in the cells. BMP2 and BMP7 have both been identified as driving gliogenesis from the adult progenitor population, and both are present at high levels in the normal ventricular wall. We predicted that if we used noggin to knock down those BMPs, we would be able to remove the gliogenic influence and allow the cells to remain in a state of undifferentiated expansion or at least responsive to the BDNF signal. We injected both noggin and BDNF. We scored the numbers of newly generated neurons at 3 weeks and 2 months and this had risen to 400 neurons/ mm^3. The noggin succeeded in potentiating the BDNF effect. This was enough in terms of numbers to justify going to a disease model, the Huntington's disease mouse, R6/2. This has a huge 150 CAG repeat expansion in the first exon of the Huntington gene, and these mice get sick by 6–8 weeks and are dead by 12. In the disease model would mobilization of the progenitor pool accomplish striatal neurogenesis? We looked at a variety of different markers of medium spiny neurons. We found an increased number of replaced striatal neurons — sufficient to justify a behavioural study. But first we needed to know whether or not the cells are connecting up, something that has been hard to demonstrate in any model. The striatum provides a nice model for this because the medium spiny cells send fibres out to the globus pallidus, almost exclusively. By putting a retrograde tracer into the globus pallidus we could see whether or not the newly generated striatal neurons actually sent fibres out to and connected up with their distal targets. We injected fluorogold into the globus pallidus and found that it traced back to the BrDU-labelled cells from the striatum. This took several weeks after injection. We used a rotorod motor function test on the injected mice, and after five or six weeks we began to see behavioural improvement with a significant difference from the controls, together with a survival advantage. This work demonstrated the ability of endogenous stem cells to subserve some useful function.

Weissman: In most of the experiments, certainly the ones I have been involved with and yours, there is usually an intermediate step of growth *in vitro* and the cells are then put back *in vivo*. Then you could show that a single cell-induced neurosphere gives at least trilineage differentiation. We do it by prospective isolation with surface markers. I don't think we have ever actually compared our surface $CD133^{+}24^{-}$ with your neural progenitor, but I wouldn't be surprised if it was the same. The experiment that hasn't been done is to look at whether they are neural stem cells without adding FGF, LIF or EGF. Do you induce a state that is not normally there? We need good clonogenic marking of cells, because the ultimate definition of a stem cell is as a single cell that differentiates to all lineages. I haven't seen this in the neural system yet: have you?

Goldman: Not really, but that is tough to demonstrate. The closest we have come to this is the use of prenatal intraventricular grafting of neural stem cells. In our work with the oligodendrocyte and white matter progenitors, we looked at the lineage potential of these cells in E17 fetal rat brains, by using intraventricular grafts delivered in one of two ways. One was to culture the cells for a week, expand them and inject. The other was to inject freshly sorted cells straight into the brain. In both cases, we saw all major neural lineages generated, both neurons and glia.

Björklund: I would like to get back to your white matter-derived progenitor cells. Have your tried to address whether they can be expanded for a long time and whether they are clonogenic?

Goldman: They are clonogenic but they can't be expanded for a long period. We have taken them to three or four months at which point the line dies out. This has been uniform across all samples. We have looked at the telomerase level, which is not detectable in the cells at the time of initial harvest. We think the cells are undergoing normal replicative senescence.

Björklund: You don't think that the initially isolated white matter progenitor is gradually lost during repeated passages? If so, your cultures would become enriched with more committed precursors over time.

Goldman: We have tried to control for this with repetitive rapid passage. This didn't help. We have tried to overexpress telomerase, which has been successful in other lineages, but it didn't work here.

Grompe: Do you need brain biopsies for this?

Goldman: These are surgical resections.

Grompe: What about postmortem brain specimens?

Goldman: We looked at this years ago but the numbers of viable progenitors isolated from cadaveric brain were so low that it wasn't worth the trouble. Rusty Gage's lab published a paper a couple of years ago (Palmer et al 2001) advocating this, but just as we'd found, the numbers they achieved were quite low.

Minger: You get 99% astrocytes and very few neurons.

Grompe: You haven't isolated them with the same procedure. Basically, it is not known at this point.

Goldman: That's right: we never went back to the autopsy brain to utilize the same technique as Rusty's.

Grompe: If your success here is near 100%, in terms of the donor pool, it would be worthwhile asking the question as to whether, for at least a couple of hours after death, these cells are still around. You could get a lot of tissue this way.

Goldman: For allografts we are better off using fetal material. We can purify the same cells from late-stage fetuses in numbers that are much greater than those here.

Nadal-Ginard: Where are these cells coming from?

Goldman: This is a good question: we know nothing about the ontogeny of these cells. Developmentally there is a glial progenitor population that emigrates from the subventricular zone and then resides within. Which of these cells undergo local clone expansion to generate astrocytes and which remain as progenitors, we don't know.

Evans: Do you hypothesize that these cells are present on all the neural ependyma, or is it just in these two specific zones?

Goldman: I believe that they are present throughout the ependymal areas. We have looked specifically along just the lateral ventricular subependyma, and they are present throughout the entire wall of the lateral ventricle. This is in rats and mice by functional criteria and growth criteria, and in humans by sampling and histological criteria. Others have reported that neural stem cells can be derived as such from the third and fourth ventricle as well as the spinal canal in the adult rat. We have not been able to replicate the spinal canal results.

Evans: Are those neural stem cell isolations the same cells as the ones that you are talking about?

Goldman: Not necessarily. Weiss's technique of neurosphere identification of neural stem cells involves long-term expansion and essentially selective emergence of stem cell clones within a larger population. We are directly sorting out the progenitor population.

Björklund: Can you get cells with these characteristics directly from the resected white matter by standard proliferative techniques, without the cell sorting step?

Goldman: If we just take the unsorted white matter and grow it under the same conditions we see plenty of what appears to be neurosphere generation. Actually, there is a higher incidence because we are not losing cells via the sorting process. But when we look at the proportion of those that are trilineage it is much lower. We are losing the multipotential clones. These cells are growing in mixed culture, with a very high astrocytic predominance.

Björklund: It is often stated that cells with stem cell-like characteristics in the adult brain can only be isolated from the subventricular zone. This would imply that if you use the same isolation technique on tissues remote from the subventricular zone you wouldn't be able to expand these types of cells. How would you see this?

Goldman: A correct phrasing might be that you can't make that determination until you look perhaps four months after you have cultured the material. At this point the ventricular zone population will still be going strong and the white matter population will be history.

Björklund: This subpopulation would then fit with some kind of immature progenitor but with an unclear capacity for self-renewal.

Goldman: I am trying to distinguish here between a transit amplifier as opposed to a truly self-renewing pool. Then, of course, the issue is whether or not one can revert back to the other. In many other tissues it can: here it is not so clear.

References

Palmer TD, Schwartz PH, Taupin P, Kaspar B, Stein SA, Gage FH 2001 Cell culture. Progenitor cells from human brain after death. Nature 411:42–43
Rakic P 2004 Neuroscience: immigration denied. Nature 427:685–686

General discussion I

Stem cells in the brain

Gearhart: I have a general question. We have transplanted several thousand animals with neural stem cells derived from human embryonic germ (EG) cells, including rats, mice and monkeys. We have found a high percentage of cell death following grafting, extensive cell migration, and that many cells don't appear to differentiate following grafting. The sense is that we have a lot of obstacles here to make this work in a safe manner. How are we going to manage these different problems? Will it eventually work?

Goldman: That cuts to the core issue: therapeutically, what is the best cell population to use, giving the best balance of efficacy and safety? I am biased in that safety has to be predominant and I feel that tissue-derived cells will be best to work with until we understand more about the embryonic stem (ES) and EG-derived cells. We don't have to worry then about teratomas.

Gearhart: With our human cells we have never seen a tumour. I am concerned that we see what are apparently undifferentiated cells and cells migrating widely in the CNS grafts. The other thing is that in some of the animals we are seeing nice improvements, so we are encouraged on one hand and on the other we see undifferentiated cells.

Goldman: Improvement cuts to the issue of efficacy but not safety. It is the undifferentiated cells we have to worry about, not just in terms of teratoma formation: heterotopic neuroma formation in white matter would be even worse.

Minger: Part of the problem is going into adult brain. Adult brain is a wasteland! If you are trying to transplant neural stem cells into adult brain, unless it has been lesioned you get almost no differentiation at all. We have done this routinely. We have put in forebrain progenitor cells expanded for three or four weeks and they survive nicely, but sit there as a giant pile of nestin-positive cells. With Steve McMahon at Kings College we have put fetal spinal-cord cells into an extensive cord lesion: we get extensive migration and 50% of the cells are nestin positive, and 30% differentiate into astrocytes. It is going to need combinatorial therapy rather than just a stem cell therapy. Combinations of cells and perhaps matrices, something to block glial scar formation and growth factors will be needed.

Goldman: I agree, although it depends on the disease. The adult brain is a tough target.

Björklund: With the standard neurosphere expansion procedure, in adult brains the vast majority of cells stay where they are placed. There is a subfraction that does migrate, but again the vast majority of those that migrate do not differentiate and many may die after some time. One is faced with a rather straightforward problem: in the adult brain one has to match the kind of cell, and the commitment of the cell, to the properties of the environment in which it is placed. Damaged or disease-affected brain tissue is likely to be significantly different from the intact one. Another important factor is the inflammatory response, which must be taken into consideration. This may have both positive and negative influences: it may support cell survival, but it is also possible that it could kill the cells or prevent them from differentiating.

Trounson: I thought Pluchino et al (2003) had shown in the multiple sclerosis (MS) mouse model that neural precursor cells injected into the peripheral circulation migrate into the brain, forming oligodendrocytes and neurons, and reverse the MS phenotype.

Björklund: The more telling integration was in the affected areas where there was disruption. This may open up for the cells to integrate. Whether or not cells integrate in the intact animal needs to be addressed.

Grompe: It is not the cells, but the engraftment environment that seems to matter most. Our experience in the liver is that with a proper microenvironment you can use just about any cell and get liver reconstitution. The brain will be extremely tough in that it is not a self-organizing regenerating tissue like many other parenchymal organs. It is hard to think about how you would create a selective advantage or self-organizing environment. It seems to me that the cells are not the key to this.

Minger: I would disagree with your comment that the cells aren't important. We know from the work by Professors Björklund and Gage that the cells are important: you can take cells from two different brain regions, implant them in the same site and get functional efficacy with one cell population and not the other.

Grompe: I am not really saying that they are not important, but the cells might be the more easily tractable scientific problem.

Weissman: Steve Goldman, I think I heard you say that you might be 'dedifferentiating' transit-amplifying cells. Just in case you did say that, now that we have defined all the stages in haematopoeisis we have tried to take any cell back even one step, and they don't do it. Our pathologists tell us that dedifferentiation happens, but every time we look at this carefully we cannot find it.

Goldman: No, in fact we looked for reversion to a stem cell state from transit-amplifying glial progenitors, but did not get a straightforward answer. Basically, we were trying to maintain the white matter progenitor as the multipotential stem cell, precisely to see if the white matter progenitor was able to yield neurons as well as glia while expanding itself in a self-renewing fashion. But we did not see this.

Instead, the cells were multipotential, but not self-renewing; they made neurons and glia, but either stopped expanding or simply died out after several months in culture. Arturo Alvarez-Buylla had a paper reporting that the neuronal progenitor cell, derived from ventricular zone stem cells in the presence of long-term epidermal growth factor (EGF), could revert back to a stem cell phenotype (Doetsch et al 2002).

Weissman: The problem is that he defines a cell as being an astrocyte because it has GFAP. GFAP is simply a marker, and it might be a marker of neural stem cells in the subventricular zone. They may not be astrocytes by class or definition. We went through an immunization, rat anti-mouse, to raise monoclonal antibodies to see the diversity of astrocytes. It is spectacular. If we look at some, we will only see cells in the subventricular zone; with others we will see another part. They are not a single population. To go from just GFAP is going to be a mistake. Normally, of course, in adult brain we don't see much migration. But if the brain is injured, for example by means of the middle cerebral artery ligation, which leads to extensive and almost uncontrollable levels of death, implanted neurospheres derived from neural stem cells just scream towards there. This tells us that there are going to be signals, and receptors for these signals—they might even be cell-type and cell-stage specific. We have a long journey ahead of us. We are just starting in this field. A lot of the diseases that we want to treat will require global migration.

Gepstein: Are there any functional assays to see whether the neurons that degenerate integrate into the existing neuron, either *ex vivo* or *in vivo*?

Goldman: This was essentially the bottom line on the behavioural assay with the Huntington mice.

Gepstein: I mean to look at whether they interact with the existing cells, transmitting information in an appropriate way.

Goldman: This has to be done at the level of electrophysiological analysis, either *in vivo* or in slice preparations with marked cells. We have not done this at that level. Essentially we are using the improvement in the behaviour to infer the underlying changes that we assume must have taken place: the newly generated cells must be sending out fibres that synaptically connect up with their targets. It is of course possible that there is a non-synaptic interaction resulting in a behavioural recovery, but this is unlikely.

Björklund: Transplants of fetal brain tissue have been analysed in some detail, demonstrating quite good integration of these cells, either as neuroblasts or young post-mitotic neurons. These kinds of cells are able to grow axons and make synaptic connections with the host neurons, as well as receiving inputs from host neurons, and restoring purposeful behaviour. The establishment of functional connectivity is in most cases highly dependent on the type of lesion used. There is little connectivity if the cells are implanted into the intact brain, but if a lesion is made prior to implantation abundant connections may form. In his recent *Nature* News and Views article Pasko Rakic suggests that even if it happens in rodents it is not going to happen in humans (Rakic 2004). This is an issue that we know less about; what we can be quite sure about is that in rats and mice repair due to neuronal integration and connectivity is quite well demonstrated. Whether these phenomena also occur in the human brain is still unclear. In humans, the only type of information we have is from cell transplantation in patients with Parkinson's disease: implanted dopamine neurons establish functional synaptic contacts with neurons in the host brain. But other types of cell implants have not been studied in this way.

Trounson: Some of the studies where stem cells are transplanted into the spinal column suggest that there is a strong trophic factor response which may induce autologous repair but is not related to the formation of new neurons. Does this also happen in the brain? I am referring to the kind of experiments done at Johns Hopkins, where embryonic and germ stem derivatives were transplanted into the spinal column of rodents, returning motor activity in the animals. By simply measuring behaviour, the impression is that the injected stem cells recolonized the damaged spinal column.

Björklund: The way to explain this, I think, is to postulate that there is an influence of the graft on the host brain, in the absence of any connectivity. This is usually referred to as a 'trophic' action that results in functional improvement or recovery, i.e. an influence that helps the brain to recover its own function. Whether this is due to glial cells exerting trophic influences, or whether it is some other type of effect — for example, in counteracting degenerative changes or preventing the damage from becoming worse — isn't well understood. Such effects may explain why functional recovery is seen in spinal-cord lesion models with stem cell implants that don't seem to differentiate into functional neurons, or establish any functional connectivity.

Solter: Presumably you will be able to recover the simple sensory or motor function by replacing the missing neuron, no matter what kind of neuron does the replacing. But one question is, if someone with Alzheimer's forgot the Iliad in Greek which they once knew, will they now remember it when the cortical neurons are replaced? Are the cortical neurons going to go to the right place and recreate the right, pre-existing synapses?

Weissman: Well at least they might be able to learn it again! I want to pick up on what Anders Björklund was just staying. In the kinds of lesions that are seen in crush injury of the spinal cord rather than transaction, or localized injury in the brain, how often do you see demyelination as a primary issue and how long can axons live without myelination?

Goldman: We need to distinguish between the experimental scenario and the clinical syndrome. The model is crush injury. Clinically crush injuries are extremely rare, and these come from traumas such as displacement of vertebral bodies and fracture. Transection is also rare. Contusion injuries are more common, and these are caused by flexion–extension injuries that demyelinate several segments at a time, often with infarct. The deep penetrating arteries into the spine are essentially getting squashed. The primary concussive injuries are a subset of the flexion–extension injuries, and are still largely demyelinative. In fact, the most common feature of clinically common spinal injuries is demyelination.

Weissman: There you are not necessarily looking for neurons. You are looking for better remyelination. How long can your axons survive without myelination? How long have you got?

Goldman: No one knows. The best experimental models are of central demyelination. Clinically, however long you have, it appears it is not enough because people don't improve after those injuries. When someone has had a complete segmental cord injury, when the radiography argues that the patient is demyelinated at that locus, then the ball game is over. It is not like MS where demyelination is then associated with remyelination and functional improvement. The problem is that spinal injury is at least secondarily ischaemic; the axons are deprived of bloodflow and go on to infarct, just as the local glial cells do.

Weissman: But the axons are connected to cells that are connected to bloodflow.

Goldman: It is the same set of issues as in chronic stroke. Over how long a distance can one have an acutely demyelinating injury that is on an ischaemic basis and still have an axon preserved? In the cord there is nothing in the clinical literature to answer this. The information is on subcortical stroke. Typically infarcts over a centimeter are associated with significant functional deficit.

Weissman: If you were successful in remyelinating, are there now MRI methods to look specifically for remyelination, and does the remyelination itself give you some target?

Goldman: The remyelination is clearly visible as long as you are comparing it with a demyelinating event. The second cause of demyelination is not the acute ischaemic event but rather the later venostasis from local oedema. This cuts off venous return and gives a secondary ischaemic syndrome typically a few days after the primary insult. This is also an infarct. This is what steroids are intended to prevent. If there has not been underlying infarct and the axons are left with oligodendrocyte loss and nothing else, this is the realistic clinical target. MS is the better model here where there is discrete loss of myelin in given foci and there is some degree of axonal transaction.

Weissman: What I worry about in MS is the ongoing autoimmunity. Even if you get the right cells in this will be a problem.

Goldman: I am not talking about MS as a therapeutic target, but just as an aid to predict how long a focus can be demyelinated without the axons dying. No one can give a solid answer clinically because the study hasn't been done. Experimentally though, this has been done a few ways. From our own data with a model that allows discrete demyelination, we will see axons survive that can be remyelinated for about three weeks. By six weeks there is substantial axonal loss. This has implications for the donor cells that might best be used for this purpose, because some remyelinate more quickly than others.

Weissman: How much does the formation of the syrinx affect this? Sometimes a hollow cavity is developed in the nerve cord by expansion of fluid. Does this also contribute to demyelination?

Goldman: Syrinx formation is a real clinical problem but it is very late, sometimes occurring years after the first injury.

Björklund: It is a problem for chronic spinal cord injury patients. This is another potential target for cell therapy.

References

Doetsch F, Petreanu L, Caille I, Garcia-Verdugo JM, Alvarez-Buylla A 2002 EGF converts transit-amplifying neurogenic precursors in the adult brain into multipotent stem cells. Neuron 36:1021–1034

Pluchino S, Quattrini A, Brambilla E et al 2003 Injection of adult neurospheres induces recovery in a chronic model of multiple sclerosis. Nature 422:688–694

Rakic P 2004 Neuroscience: immigration denied. Nature 427:685–686

Embryonic stem cells: a perspective

Martin Evans

Cardiff School of Biosciences, Cardiff University, PO Box 911, Cardiff CF10 3US, UK

Abstract. Embryonic stem (ES) cells are proliferating tissue-culture cells which are embryonic in origin and retain the full lineage and differentiative capacity of the early mammalian embryo. Only mouse ES cells have been fully characterized but similar cultures have been isolated from numbers of other mammalian embryos including, notably, human. ES cells can differentiate either *in vivo* in the context of an embryo or equivalently *in vitro*. Their ability to undergo genetic alteration and selection *in vitro* followed by embryonic development *in vivo* has led to their extensive use as a vector for mammalian genetic engineering. This is not an application for human ES cells but their ability to differentiate *in vitro* into a plethora of specific tissue precursors has led to scenarios for their use in cellular and tissue engineering therapies. Ethical and practical problems for the provision of *ad hominem* human ES cells remain. Knowledge from the mouse system may aid our understanding of the relationship between ES cells and embryos.

2005 Stem cells: nuclear reprogramming and therapeutic applications. Wiley, Chichester (Novartis Foundation Symposium 265) p 98–106

It is self-evident that at the earliest stages of development there will be dividing populations of cells which have a wide range of potential developmental fates. Experimental embryology has shown that fate becomes progressively restricted as a cell lineage develops. It is not self-evident at which, if any, stages along this progression from totipotential to single lineage restriction, the cells may cycle mitotically keeping that developmental fate statically intact. Indeed this question is intimately bound up with our understanding of the stability of cell determination and maintenance of the differentiated state. A precursor population is not the same as this stem cell population. In addition stem cells, both during development and, in the adult organism, reside in specific conditions, in niches, and it is not necessarily straightforward to maintain or replicate the niche conditions in isolation. In order to isolate and maintain stem cells such conditions need to be provided or their necessity obviated by other means. Another factor is that in some adult conditions stem cells may represent a very slowly growing reserve population which gives rise to transit committed populations in which the amplification of rapid cell division occurs.

We are, therefore, fortunate to have at least one example (mouse) where the early embryoblasts may be isolated into relatively simple tissue culture conditions and will divide actively whilst retaining their developmental capacity. These are cells we know as mouse embryonic stem (ES) cells.

Mouse ES cells are often described as pluripotent rather than totipotent largely through the caution of the experimental scientist. These cultured embryoblasts may, under optimal conditions, be essentially totipotent. This is most clearly demonstrated by the experiments of Nagy et al (1993) who showed that only the addition of tetraploid morula cells as carriers was sufficient to allow full development of the complete mouse from cultured ES cells. The tetraploid carrier embryo cells give rise only to extra-embryonic tissues, in particular to trophectoderm and embryonic endoderm. Embryonic endoderm is very readily differentiated from ES cells. There has, however, been some discussion about the capacity for differentiation to trophoblast. Observations of sporadic appearance in culture of trophoblast-like cells have raised questions about their clonal origin and functional identity. Current understanding of the role of *Oct4* in ES maintenance (Pan et al 2002) and the elegant experiments of (Niwa et al 2000) who modulated its expression are evidence that ES cells can indeed differentiate into trophectoderm. Rossant (2001) has provided a detailed review of the interrelationships between embryonic and trophoblast stem cells.

Mouse ES cells are therefore totipotent. Putative ES cells from other mammalian species have not been satisfactorily tested in similar conditions although bovine cells have been aggregated with tetraploid embryos and live calves born (Iwasaki et al 2000). Most surprisingly only a low level of chimerism from the ES component was reported bringing into question the origin of the remainder of the tissues of these animals.

The isolation of mouse ES cells has been well-described and we have previously discussed their tissue of origin—the embryo (Evans & Hunter 2002). The originally described (and still optimal) method for isolation was from implantationally-delayed blastocysts that were subsequently explanted into tissue culture (Evans & Kaufman 1981). It is also possible to derive them from explanted morulae (Eistetter 1989), whole blastocysts or isolated inner cell masses (ICMs) (Martin 1981). The ICM from a peri-implantational 4.5 day embryo is an optimal source if it is separated from the nascent endoderm (Brook & Gardner 1997). It would seem most likely that the source of ES cells and their homology within in the embryo is represented by the primary epiblast at an early post-implantation stage. They are probably not directly homologous to ICM cells and statements to this effect are an oversimplification. There is continuing speculation and discussion of mouse strain dependency. Initially we demonstrated isolation from a number of inbred and outbred mouse stocks. There has been subsequent isolation from a number of potentially recalcitrant strains (CBA, C3H, C57) and no particular genetic factors have been defined.

Mouse ES cells require growth conditions and growth factors which can be supplied by primary fibroblasts (mouse or chick) and which can be replaced by specific growth factors in carefully selected sera. Some progress has been made in the use of serum-free media (Horii et al 2003, Wiles & Johansson 1999, Amit et al 2004). One of the major components provided by the feeder cells is leukaemia inhibitory factor which activates the leukaemia inhibitory factor (LIF)/Stat-3 signalling pathway, and which is known to be the single factor involved in mouse ES cell maintenance (Smith 2001). Interestingly, although human ES cell lines and those from other species need a feeder layer for isolation and maintenance, there is no evidence of the action of or requirement for LIF. Recently, Sato et al (2004) have dissected the maintenance of human and mouse ES cells and demonstrated that continued stem cell self-renewal depends upon the up-regulation of the Wnt pathway. By inhibiting GSK-3 function pharmacologically they were able to maintain both mouse and human ES cells in a pluripotential condition without additional factors. This could be a most important practical development.

It is worthwhile reiterating that putative ES cells have been isolated from numbers of other mammalian and indeed non-mammalian species. One notable exception to this success has so far been the rat. In no case has it yet been possible to prove the full ES status of these other mammalian cells and their pluripotency tests rest upon *in vitro* differentiation and teratocarcinoma formation in immunosuppressed mice. From the chick, however, cells have been isolated, tissue cultured and used to make germline chimeras. These cells are in many ways similar to mouse ES cells and, interestingly, are responsive to LIF (Etches et al 1996, Pain et al 1996).

The major use of mouse ES cells has been genetic engineering using their availability and accessibility in tissue culture to prescreen for the genetic change required, be it induced in a random or targeted fashion. Clearly germline genetic manipulation in humans is undesirable but it is conceivable that techniques of targeted mutation and transgenesis might be used as a precursor to somatic cellular therapy. The main potential use seen for human ES cells is, however, in their pluripotency and ability to differentiate into a wide range of cell types *in vitro*. Hopefully these will include tissue-specific stem cells and precursors. Here it is useful to look at the information on *in vitro* differentiation gained from the mouse ES cell system.

ES cells differentiate very readily *in vitro*, typically into a wide diversity of cell types in a relatively disorganized manner. This apparent disorganization hides local organized interaction. When it was first observed that embryonic carcinoma (EC) cell differentiation proceeded via the formation of embryoid bodies, it became clear that this differentiation was not random or stochastic and the cells were not just reverting from a malignant state but following and becoming entrained in the normal developmental pathway of an early embryo. In every case of spontaneous differentiation studied it was found that the first product was always embryonic endoderm. Despite the successes of using retinoic acid and other inducers to influence the quantitative outcome it remains the case that obtaining differentiation in a specific direction depends as much upon selection of the required cell type as it does on specific induction. One exception might be the very ready differentiation of neural precursors as a possible default pathway when the cells are in isolation (Ying & Smith 2003).

There have, not withstanding my remarks above, been some remarkable successes in specific differentiation of mouse ES cells *in vitro*. McKay's group (Lumelsky et al 2001) generated clusters of cells expressing insulin and other pancreatic endocrine hormones. Glucose triggered insulin release from these cell clusters and when injected into diabetic mice, the insulin-producing cells undergo rapid vascularization and maintain a clustered, islet-like organization. This has triggered a great deal of interest and Peck et al (2002) have reviewed the possibilities.

Other dramatic results of *in vitro* differentiation of ES cells include the formation of germ cells (Toyooka et al 2003) and oocytes (Hubner et al 2003). It will be important to be able to isolate quantities of the desired differentiated cell type, which so far has been done by selection (or ablation of other types). A good example is the description by Zandstra et al (2003) of their method for selection of cardiac myocytes by expression of a neo-transgene under myosin heavy chain promoter control. The cells are stimulated to differentiate in the usual context of embryoid bodies before selection. Here as in most other cases the spontaneous differentiation is exploited in an unsophisticated manner.

It is not yet clear whether other strategies might prove useful such as direct transdifferentiation. Just as we need to know how to reverse a differentiated cell to a dedifferentiated state by a direct manipulation, we also need to know how to transdifferentiate a pluripotential ES cell into the required product. One promising study comes from work with *Xenopus* animal cap cells where ectopic expression of GATA-4 can drive them directly towards cardiac muscle (Latinkic et al 2003). It may be that we will come to understand how to use cocktails of appropriate transcription factors to drive desired differentiation.

Discussion

The ideas for developing *ad hominem* tissue stem cells for therapeutic tissue repair depend upon de-differentiation of adult cells of one lineage and re-differentiation into a new lineage(s). Amplification of these cell populations via *in vitro* culture

TABLE 1

(a)

- Stages envisages for the use of ES cells
  - —Choice and source of the differentiated cell parent
  - —Dedifferentiation via oocyte and embryo
  - —Dedifferentiation via cell fusion
  - —Dedifferentiation via exogenous cell soups and factors
  - —Dedifferentiation via specific transcription factor manipulation
  - —Validation of the embryonic stem cell line:
    - genetic — karyotype
    - epigenetic — methylation pattern; imprinting
    - differentiative status

(b)

- Differentiation to a pure and appropriate population of committed precursors
  - —Culture conditions GMP; ?xenotransplant
  - —Methods for differentiation
  - —Niches
  - —Committed precursor may be more desirable than a differentiated end cell
  - —Validation of purity and specificity of differentiated population

would be advantageous and the use of ES cells remains an attractive scenario. These cells will most probably not be isolated from embryos but derived by an equivalent de-differentiation process driven directly. They will need to be re-differentiated into the desired precursor population. For both the de-differentiation and re-differentiation processes we need to understand the details of the cell and embryo processes of normal development in much greater depth than presently known. Highlighted in Tables 1a,b are the stages envisaged in the use of ES cells as a source for therapeutic tissue engineering. At every step it is clear that we will require a much better theoretical understanding of the processes as well as the practical ability to control them. This falls within the realm of experimental embryology and it is becoming clear that despite the great advances of recent years much remains to be done.

References

Amit M, Shariki C, Margulets V, Itskovitz-Eldor J 2004 Feeder layer- and serum-free culture of human embryonic stem cells. Biol Reprod 70:837–845

Brook FA, Gardner RL 1997 The origin and efficient derivation of embryonic stem cells in the mouse. Proc Natl Acad Sci USA 94:5709–5712

Eistetter HR 1989 Pluripotent embryonal stem cell lines can be established from disaggregated mouse morulae. Dev Growth Differ 31:275–282

Etches RJ, Clark ME, Toner A, Liu G, Gibbins AM 1996 Contributions to somatic and germline lineages of chicken blastodermal cells maintained in culture. Mol Reprod Dev 45:291–298

Evans MJ, Kaufman MH 1981 Establishment in culture of pluripotential cells from mouse embryos. Nature 292:154–156

Evans M, Hunter S 2002 Source and nature of embryonic stem cells. C R Biol 325:1003–1007

Horii T, Nagao Y, Tokunaga T, Imai H 2003 Serum-free culture of murine primordial germ cells and embryonic germ cells. Theriogenology 59:1257–1264

Hubner K, Fuhrmann G, Christenson LK et al 2003 Derivation of oocytes from mouse embryonic stem cells. Science 300:1251–1256

Iwasaki S, Campbell KH, Galli C, Akiyama K 2000 Production of live calves derived from embryonic stem-like cells aggregated with tetraploid embryos. Biol Reprod 62:470–475

Latinkic BV, Kotecha S, Mohun TJ 2003 Induction of cardiomyocytes by GATA4 in Xenopus ectodermal explants. Development 130:3865–3876

Lumelsky N, Blondel O, Laeng P et al 2001 Differentiation of embryonic stem cells to insulin-secreting structures similar to pancreatic islets. Science 292:1389–1394

Martin GR 1981 Isolation of a pluripotent cell line from early mouse embryos cultured in medium conditioned by teratocarcinoma stem cells. Proc Natl Acad Sci USA 78:7634–7638

Nagy A, Rossant J, Nagy R, Abramow-Newerly W, Roder JC 1993 Derivation of completely cell culture-derived mice from early-passage embryonic stem cells. Proc Natl Acad Sci USA 90:8424–8428

Niwa H, Miyazaki J, Smith AG 2000 Quantitative expression of Oct-3/4 defines differentiation, dedifferentiation or self-renewal of ES cells. Nat Genet 24:372–376

Pain B, Clark ME, Shen M et al 1996 Long-term in vitro culture and characterisation of avian embryonic stem cells with multiple morphogenetic potentialities. Development 122: 2339–2348

Pan GJ, Chang ZY, Scholer HR, Pei D 2002 Stem cell pluripotency and transcription factor Oct4. Cell Res 12:321–329

Peck AB, Cornelius JG, Chaudhari M, Shatz D, Ramiya VK 2002 Use of in vitro-generated, stem cell-derived islets to cure type 1 diabetes: how close are we? Ann NY Acad Sci 958:59–68

Rossant J 2001 Stem cells in the mammalian blastocyst. Harvey Lect 97:17–40

Sato N, Meijer L, Skaltsounis L, Greengard P, Brivanlou AH 2004 Maintenance of pluripotency in human and mouse embryonic stem cells through activation of Wnt signaling by a pharmacological GSK-3-specific inhibitor. Nat Med 10:55–63

Smith AG 2001 Embryo-derived stem cells: of mice and men. Annu Rev Cell Dev Biol 17: 435–462

Toyooka Y, Tsunekawa N, Akasu R, Noce T 2003 Embryonic stem cells can form germ cells in vitro. Proc Natl Acad Sci USA 100:11457–11462

Wiles MV, Johansson BM 1999 Embryonic stem cell development in a chemically defined medium. Exp Cell Res 247:241–248

Ying QL, Smith AG 2003 Defined conditions for neural commitment and differentiation. Methods Enzymol 365:327–341

Zandstra PW, Bauwens C, Yin T et al 2003 Scalable production of embryonic stem cell-derived cardiomyocytes. Tissue Eng 9:767–778

DISCUSSION

Jaenisch: Let's return to the question of whether ES cells correspond to anything *in vivo*. I would argue that they don't at all. I think they are an artefact of tissue culture, where there has been selection for proliferation of cells under adverse conditions. They just happen to be pluripotent. You did some crude expression profiling. If you think biologically, they are quite different to anything that occurs normally: they are certainly not ICM cells. If you look at ES cells they are not epigenetically very stable. In contrast, epiblast cells are very stable.

Evans: We don't know that.

Jaenisch: We have looked at this. Consider the imprinted genes such as H19: they change. We looked at subclones of a parental line and two daughter cells are different.

Evans: You are right. Subclones of ES cells can change, but by this time they are artefacts of culture. The question in my mind is slightly different: I want to know whether the epiblast cell in the embryo is what we are seeing? If we have a complete artefact, this could be very useful, because we could find ways of getting other complete artefacts.

Jaenisch: I think ES cells are very useful. I am just arguing that they are artefacts.

Evans: Perhaps all we need to do is to understand the artefact systems, and we could then get other stem cells in culture in the same way.

Jaenisch: ICM cells explanted in culture, as Austin Smith has shown, turn off *Oct4*, sit in the dish and do nothing. Some cells then turn on *Oct4* and begin to proliferate: therefore there is a selection for proliferation. As a side product these cells become pluripotent.

Evans: What happens with other stages? I don't think the ICM is the best stage to use. If you explant later stages in, the cells grow out straight away.

Weissman: Quite often in our role as developmental biologists we like to make claims for groups of cells when only a rare cell in that group might be the important one. The best demonstration comes from 1960–1967. This was at the time when Cliff Grobstein was doing a lot of epithelial–mesenchymal interaction studies. Robert Auerbach at Wisconsin showed that he could separate the mouse thymus into epithelial and mesenchymal components, culture them *in vitro* and reconstitute development of lymphocytes before T cells were discovered. Then he separated the endoderm from the mesenchyme across a 0.45 μm Millipore filter and found to his shock that the lymphocytes always came from the endoderm or epithelial side, not the mesenchymal side. It didn't matter whose mesenchyme he put across: it was the endoderm that counted. He concluded at the time that it was the endoderm that was the true precursor of the lymphocyte. Then Malcolm Moore and John Owen showed that in reconstructing that system and whole epithelium, there was a very rare cell staining differently. They called it a haemocytoblast, and this was the actual origin within the endoderm layer of all the cells just a few days later. They couldn't see the epithelium any more. A very rare precursor, progenitor or stem cell within a whole tissue can give rise to everything and we are just being fooled by our eyes.

Evans: I think that is right, but we are not looking at a large population of initial cells.

Weissman: When Rudi Jaenisch says that if you take cells out, all of them shut off *Oct4* and then one regains it, it makes me wonder whether there was one that kept *Oct4* on and you didn't see it. You are looking at generalized profiles: we don't have single cell sensitivity.

Evans: Yes, there could be one embryoblast in the inner cell mass. The isolated inner cell mass, though, is only 15–50 cells.

Weissman: It would be helpful if we could develop marking systems to follow step by step what is happening, tracking each cell as it goes along.

Minger: I'd like to comment on what Rudi Jaenisch said about selecting for cells. At Kings, we have had about 80 high-quality human embryos that we have attempted to make stem cells from. Of these, about 40 have a large, demonstrable inner cell mass. We have put these into culture and used the same batch of mouse feeders throughout and the same lot of serum. Susan Pickering who did this work is an incredibly good embryologist who spends a lot of time documenting every embryo that we have used. Of these 40 high quality inner cell masses we have generated just four lines. We have no idea why some ICMs generate stem cells and some don't. Almost always we get something to grow, but only in rare cases do we see a stem cell colony come up.

Evans: In the mouse we have one advantage, in that we have control over the genetics. This may be a big factor.

Minger: That is right.

Solter: But in the mouse you will still not have a 100% success rate in generating ES cells.

Evans: The best is more than 25%.

Minger: We started out with mouse blastocysts and we didn't seem to have very many problems.

Jaenisch: We have no problems getting an 80% success rate with 129 mice—with Balb/c there is less success, and with other strains even less. There is clearly a genetic component.

Wilmut: I'd like to mention a rather indirect approach to this that was carried out by David Wells when he was a student with Martin Hooper and myself. Martin and I had a coffee-time-type conversation back in the mid-1980s, and he commented that cells seemed to change during derivation. This led first to the recognition that we don't know what changes occur during derivation of ES cells. However, we hypothesized that if we stress embryos in different ways to induce epigenetic differences, we might hope to find some changes that promote ES cell derivation. The first thing we tried was heat shocking the embryos. This increased the proportion of embryos from which ES cell lines were derived by a multiple of between two- and threefold. We weren't able to go very far with this within a studentship, but some of the lines were taken through and did form germline chimeras: there is no question that these were ES cell lines. This simple treatment of 42 °C for 10 min increased the proportion in this way.

Murray: We heard this morning that the route will be through dedifferentiation and redifferentiation. If I listened correctly, I heard Irv Weissman say yesterday that he's never seen evidence of dedifferentiation. How can we reconcile these two statements?

Evans: At the moment there is only one technique we know that really gives rise to dedifferentiation, and this is the cell nuclear transplantation into an oocyte.

Weissman: Which most of us believe doesn't happen naturally!

Evans: De-differentiation of the genetic material is also seen in cell fusion, in particular the fusion of a differentiated cell with an ES cell. But then, of course, we end up with a hybrid cell. It is not a clean system. My remark about de- and redifferentiation is that I believe that this is the future. We are talking about the stability of the differentiated state. We know that, effectively, this is controlled by transcription factor networks and secondary modification of the nucleic acid and chromatin by DNA methylation and histone acetylation. These are all reversible situations.

Trounson: It is apparent that you can make ES cells in the mouse from 129 or BL/6 mice, but it is pretty difficult from the other strains. It hasn't been successful in the rat nor, despite big efforts, in ruminants.

Evans: There has been quite a lot of success in ruminants. The problem has been the cost of carrying the experiments forward *in vivo*.

Trounson: The mouse is very strain specific, and in human there is tremendous variability in the ability of researchers to make ES cells and there's also a lot of variability in the cell lines produced. Some of the worst ICMs I have ever seen have turned into beautiful ES cell lines. The association between morphology and outcome can be poor. In the mouse, what is it about the specific strains that allows ES cell generation?

Evans: We don't know. My first point would be that it is certainly not restricted to a single strain of mice. We have done it from numbers of different inbred strains and quite a number of outbreds. There are clearly factors which influence how easy it may be from any particular strain. One such factor might be the rate of development.

Nuclear cloning, epigenetic reprogramming and cellular differentiation

Rudolf Jaenisch*†, Konrad Hochedlinger* and Kevin Eggan*†

Whitehead Institute and Department of Biology†, Massachusetts Institute of Technology, 9 Cambridge Center, Cambridge, MA 02142, USA*

Abstract. The full-term development of sheep, cows, goats, pigs and mice has been achieved through the transfer of somatic cell nuclei into enucleated oocytes. Despite these successes, mammalian cloning remains an inefficient process, with a preponderance of reconstructed embryos failing at early- to mid-gestation stages of development. The small percentage of conceptuses that survive to term are characterized by a high mortality rate and frequently display grossly increased placental and birth weights. It is likely that inappropriate expression of key developmental genes may contribute to lethality of cloned embryos. One of the most interesting issues of nuclear cloning is the question of genomic reprogramming, i.e. whether successful cloning requires the resetting of epigenetic modifications which are characteristic of the adult donor nucleus. Processes such as X-inactivation and genomic imprinting are known to depend on epigenetic modifications of the genome. The classical nuclear transfer experiments with frogs have suggested that the source of the donor nucleus affects the phenotype of the clone. We have, using expression profiling, compared gene expression in clones derived from embryonic stem (ES) cells and from somatic donor cell nuclei and find substantial gene dysregulation. Our results suggest that faulty reprogramming is caused by the nuclear cloning procedure itself. In addition, the type of donor nucleus contributes to the abnormal expression pattern seen in cloned animals. One of the major unresolved issues has been whether nuclei of terminally differentiated cells can be reprogrammed by transfer into the oocyte. To address this question we have derived monoclonal mice from B and T cells and used the genetic rearrangements of the immunoglobulin and T cell receptor genes to retrospectively verify the differentiation state of the donor nucleus. Finally, we discuss our recent studies on the reprogramming of nuclei from terminally differentiated neurons and from cancer cells.

2005 Stem cells: nuclear reprogramming and therapeutic applications. Wiley, Chichester (Novartis Foundation Symposium 265) p 107–121

Almost half a century ago the technique of nuclear transplantation (NT) was pioneered in amphibians (Di Berardino et al 2003, Gurdon 1999a). These experiments demonstrated that the nuclei of somatic cells are totipotent but that the ability to generate live animals decreases with developmental age of the donor nucleus. The generation of Dolly from an adult mammary gland cell demonstrated that at least some cells within an adult organism retain totipotency and are able to direct development of a new animal (Wilmut et al 1997). After Dolly additional mammalian species were successfully cloned from somatic cells, albeit with a low efficiency as in most cases only 0.5% to 10% of reconstructed oocytes develop into apparently healthy adults (Rideout et al 2001).

Epigenetic regulation of gene expression is recognized to be one of the key mechanisms governing embryonic development and cellular differentiation, and it is of major interest to define the specific epigenetic states that distinguish the embryonic genome from that of the different somatic cells. In normal development the process of differentiation from embryonic to differentiated cells involves alterations in the epigenetic conformation of the genome such as DNA methylation or chromatin modifications (Bird 2002, Jaenisch & Bird 2003). In cloning, exposure to the egg cytoplasm can reverse this process and convert the somatic nucleus into an embryonic state. Even in the adult organism, some cells may retain the potential to differentiate into functional cells of different lineages as has been claimed in numerous studies over the course of the last few years (this process is also designated as 'transdifferentiation') (Weissman et al 2001). Thus, we have to understand the epigenetic states that distinguish the genome of embryonic cells from that of different somatic cells (see Fig. 1). To alter the phenotype of cells in a rational way we first need to define the molecular

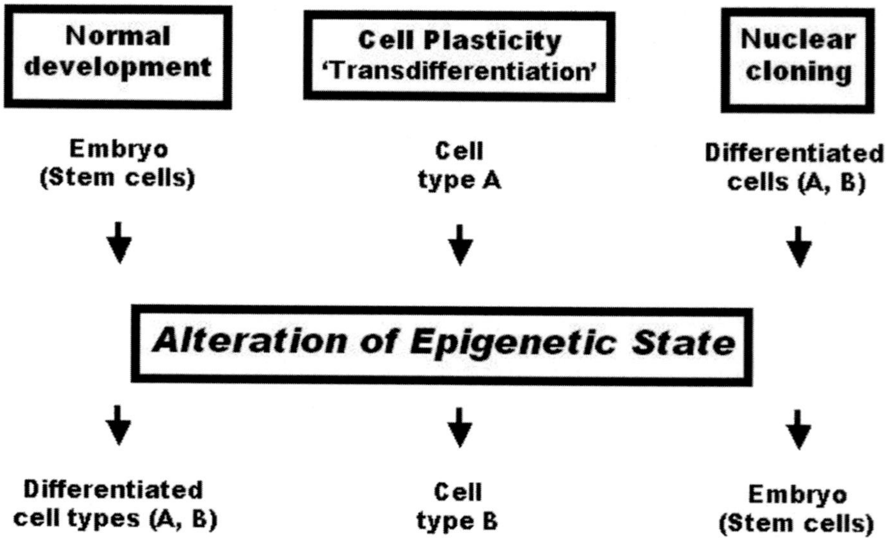

FIG. 1. Development, cellular plasticity and nuclear cloning.

parameters that distinguish these different cell types. Nuclear cloning represents an unbiased tool to begin unravelling these mechanisms as it provides a functional readout of the epigenetic changes induced by the egg cytoplasm. In this article we will summarize recent work from our laboratory that focuses on the relation between the state of differentiation and the efficiency of epigenetic reprogramming by nuclear transfer. This is relevant for the possible therapeutic applications of the nuclear transfer technology.

Epigenetic reprogramming in normal development and after nuclear transfer

The majority of cloned mammals derived by nuclear transfer (NT) die during gestation, display neonatal phenotypes resembling 'Large Offspring Syndrome' (Young et al 1998), often with respiratory and metabolic abnormalities, and have enlarged and dysfunctional placentas (Rideout et al 2001). In order for a donor nucleus to support development into a clone, it must be reprogrammed to a state compatible with embryonic development. The transferred nucleus must properly activate genes important for early embryonic development and also suppress differentiation-associated genes that had been transcribed in the original donor cell. Inadequate 'reprogramming' of the donor nucleus is thought to be the principal reason for developmental failure of clones. Since few clones survive to birth, the question remains whether survivors are normal or merely the least severely affected animals, making it to adulthood despite harbouring subtle abnormalities originating from faulty reprogramming (Rideout et al 2001).

Evidence obtained over the last few years has given insights into molecular changes that are abnormal in clones. Table 1 summarizes some of the epigenetic differences that distinguish cloned from normal animals as a result of faulty reprogramming. For the following discussion it is useful to compare the different stages of development following nuclear transplantation. The stages of development that are depicted in Table 1 are (i) gametogenesis, (ii) cleavage, (iii) post-implantation and (iv) postnatal development.

Gametogenesis

The most important epigenetic reprogramming in normal development occurs during gametogenesis, a process that renders both sperm and oocyte genomes 'epigenetically competent' (Rideout et al 2001) for subsequent fertilization. In cloning, this process is shortcut and most problems affecting the 'normalcy' of cloned animals may be due to the inadequate reprogramming of the somatic nucleus following transplantation into the egg. Since the placenta is derived from the trophectoderm lineage, which constitutes the first differentiated cell type of the

TABLE 1 Normal vs. cloned embryos

Stage	*Normal embryos*	*Cloned embryos*
Gametogenesis	Genome 'competent' for activation of 'early' genes, establishment of imprints	*None*
Cleavage	Global demethylation Activation of embryonic ('Oct4-like') genes	*Abnormal methylation* *Stochastic/faulty activation of 'Oct4-like' genes*
Post-implantation	Global *de novo* methylation, X-inactivation, telomere length adjustment (postzygotic events)	
	Normal imprinting and gene expression	*Abnormal imprinting, global gene dysregulation*
Postnatal	Normal animal	*Larger offspring syndrome, premature death, etc.*

embryo, one might speculate that reprogramming and differentiation into this early lineage is compromised in most cloned animals. Indeed, results obtained in our laboratory and by others indicate that the fraction of abnormally expressed genes in cloned newborns is substantially higher in the placenta as compared to somatic tissues (Fulka et al 2004, Humpherys et al 2002). In contrast to epigenetic reprogramming that occurs pre-zygotically, it appears that post-zygotic reprogramming such as X-chromosome inactivation (Eggan et al 2000) and telomere length adjustment (Betts et al 2001, Lanza et al 2000, Tian et al 2000, Wakayama et al 2000) are faithfully accomplished after nuclear transfer and, therefore, would not be expected to impair survival of cloned animals.

Cleavage

During cleavage, a wave of genomewide demethylation removes the epigenetic modification present in the zygote so that the DNA of the blastocyst is largely devoid of methylation. Between implantation and gastrulation, a wave of global *de novo* methylation re-establishes the overall methylation pattern, which is then maintained throughout life in the somatic cells of the animal (Jaenisch 1997, Reik et al 2001). In cloned embryos abnormal methylation at repetitive sequences (Kang et al 2001) and frequent failure to reactivate *Fgf4*, *Fgf2r* and *IL6* (Daniels et al 2000) have been observed. To investigate gene expression, the activity of 'pluripotency genes' such as *Oct4* that are silent in somatic cells but active in embryonic cells was examined in cloned embryos. Strikingly, the reactivation of these '*Oct-4-like*' genes was shown to be faulty and random in somatic clones

(Boiani et al 2002, Bortvin et al 2003). Because embryos lacking *Oct-4* arrest early in development, incomplete reactivation of *Oct-4*-like genes in clones might be the cause of the frequent failure of the great majority of NT embryos to survive the post-implantation period. To what extent the epigenetic modification—of chromatin structure and DNA methylation, which occurs in normal development—needs to be mimicked for nuclear cloning to succeed remains an unresolved question. However, the available evidence is entirely consistent with faulty epigenetic reprogramming causing the abnormal gene expression in cloned animals.

Post-implantation

The most extensive analysis of gene expression has been performed in newborn cloned mice. Expression profiling showed that 4–5% of the genome and between 30 and 50% of imprinted genes are abnormally expressed in placentas of newborn cloned mice (Humpherys et al 2002). This argues that mammalian development is surprisingly tolerant of widespread gene dysregulation and that compensatory mechanisms assure survival of some clones to birth. However, the results suggest that even surviving clones may have subtle defects that, though not severe enough to jeopardize immediate survival, will cause an abnormal phenotype at a later age.

Postnatal development

The generation of adult and seemingly healthy adult cloned animals has been taken as evidence that nuclear transfer can generate normal cloned animals, albeit with low efficiency. Indeed, a routine physical and clinical laboratory examination of 24 cloned cows of 1–4 years of age failed to reveal major abnormalities (Lanza et al 2001). Cloned mice of a corresponding age as that of the cloned cows (2–6 months in mice vs. 1–4 years in cows) also appear 'normal' by superficial inspection. However, when cloned mice were aged, serious problems, not apparent at younger ages, became manifest. One study found that the great majority of cloned mice died significantly earlier than normal mice, succumbing with immune deficiency and serious pathological alterations in multiple organs (Ogonuki et al 2002). Another study found that aged cloned mice became overweight with major metabolic disturbances (Tamashiro et al 2002). Thus, serious abnormalities in cloned animals may often become manifest only when the animals age. In summary, the available evidence suggests that truly normal clones may be the exception (Jaenisch 2004).

Derivation of cloned animals from terminally differentiated cells

A question already raised in the seminal cloning experiments with amphibians was whether terminal differentiation would diminish the potency of a nucleus to direct development after transfer into the oocyte (Di Berardino 1980, Gurdon 1999b). The recent isolation of rare adult stem cells from somatic tissues and reports of their developmental plasticity raise an important question: do viable clones only result from adult stem cells randomly selected from the donor cell population (Hochedlinger & Jaenisch 2002a)? These cells might be similar to embryonic stem (ES) cells that require less reprogramming and support post-implantation development with high efficiency. In order to better define the influence of the donor nucleus on the development of cloned animals, we have compared cells of different developmental stages for their potency to serve as nuclear donors.

Table 2 summarizes the potential of blastocysts derived from normal zygotes with that derived from ES cell and somatic donor cell nuclei after transfer into oocytes. The results show that ES NT embryos develop to term at a 10–20-fold higher efficiency than embryos from cumulus or fibroblast donor cells. The main conclusion that can be drawn from the observations summarized in Table 1 is that the nucleus of an undifferentiated embryonic cell is more amenable to, or requires less reprogramming, than the nucleus of a differentiated somatic cell. The epigenetic (or pluripotent) state of the genome in an ES cell may more closely resemble that of the early embryo, which enables ES cells to serve as more effective nuclear donors. This notion was directly tested by analysing the expression of genes that are active in the different donor cells with the expression pattern seen in cloned blastocysts derived from the respective donor cells (see above).

The results summarized in Table 2 suggest that the nucleus of an undifferentiated embryonic cell might be more amenable to, or require less

TABLE 2 Higher survival of mice cloned from ES cell donor nuclei as compared to somatic donor cell nuclei

Donor cells	*Survival to adults (from cloned blastocysts)*	*References*
Cumulus	1–3%	Wakayama & Yanagimachi 1999, 2001
Fibroblasts	0.5–1%	
Sertoli cells	1–3%	
B, T cells, neurons	<0.001%	Hochedlinger & Jaenisch 2002a, Eggan et al 2004
ES cells	15–25%	Eggan et al 2001

reprogramming, than the nucleus of a differentiated somatic cell. The epigenetic (or pluripotent) state of the genome in an ES cell may more closely resemble that of the early embryo, which enables ES cells to serve as more effective nuclear donors. Assuming that the epigenetic state of somatic stem cells resembles that of ES cells it appears possible that most surviving clones could be derived from nuclei of rare somatic stem cells present in the heterogeneous donor cell population rather than from nuclei of differentiated somatic cells as has been assumed (Hochedlinger & Jaenisch 2002a, Liu 2001). Because in previous experiments no unambiguous marker had been used that would retrospectively identify the donor nucleus, this possibility could not be excluded. Thus, it had not been resolved whether the genome of a truly terminally differentiated cell could be reprogrammed to an embryonic state. To address this issue, we used nuclei from mature B and T cells and from terminally differentiated neurons as donors to generate cloned mice.

Monoclonal mice from mature immune cells

The monoclonal mice were generated from nuclei of peripheral lymphocytes where the genetic rearrangements of the immunoglobulin and TCR genes could be used as stable markers revealing the identity and differentiation state of the donor nucleus of a given clone. Because previous attempts to generate monoclonal mice had been unsuccessful, we used a two step cloning procedure by producing first ES cells from cloned blastocysts and in a second step monoclonal mice by tetraploid-embryo complementation (Eggan et al 2001, Nagy et al 1993). Animals generated from a B or T cell donor nucleus were viable and carried fully rearranged immunoglobulin or T cell receptor genes in all tissues (Hochedlinger & Jaenisch 2002b). Our results constituted the first unequivocal demonstration of nuclear cloning from two terminally differentiated donor cell-types. The frequency of directly deriving cloned embryos from mature B and T cells, while difficult to estimate, is likely significantly lower than that of deriving clones from fibroblasts or cumulus cells (possibly less than 1 in 3000 operated embryos; Table 1). This is consistent with the notion that genomic reprogramming of a terminally differentiated cell may be extremely inefficient.

Cloned mice from mature olfactory neurons

Nuclei from post-mitotic cells that have irreversibly exited the cell cycle as part of their programme of differentiation have not been demonstrated to retain the capacity to direct embryogenesis in mammalian cloning experiments. These considerations have led to the suggestion that post-mitotic cells might be refractory to epigenetic reprogramming or alternatively might have acquired changes in the DNA that could limit their developmental potential (Rehen et al

2001). Consistent with this notion are previous experiments that failed to generate live mice from neurons (Yamazaki et al 2001). This led to the suggestion that the DNA of post-mitotic neurons might undergo rearrangements to generate neural diversity and that these changes in DNA sequence may prevent these nuclei from re-entering the cell cycle and directing embryogenesis (Vassar et al 1993, Zhang & Firestein 2002). One particularly clear example of neuronal diversity is provided by the olfactory sensory epithelium. In the mouse, each of the two million cells in the olfactory epithelium expresses only one of ~1500 odorant receptor genes such that the functional identity of a neuron is defined by the nature of the receptor it expresses. The pattern of receptor expression is apparently random within one of four zones in the epithelium suggesting that the choice of receptor gene may be stochastic. One mechanism to permit the stochastic choice of a single receptor could involve DNA rearrangements (Chun & Schatz 1999).

We have generated fertile adult mouse clones by transferring the nuclei of post-mitotic olfactory neurons into enucleated oocytes (Eggan et al 2004). In a similar approach as used for the generation of the monoclonal mice, in a first step ES cells were derived from the cloned blastocyst and cloned mice were derived subsequently by tetraploid complementation. As summarized in Table 1, the efficiency of deriving cloned ES cells from olfactory neurons was in the same range as that for nuclei from immune cells. These observations indicate that a post-mitotic neuronal nucleus can re-enter the cell cycle and can be reprogrammed to pluripotency. The DNA of mice derived from sensory neurons revealed no evidence for rearrangements of the expressed olfactory receptor gene. In addition, the pattern of receptor expression in these mice was indistinguishable from wild type animals, indicating that irreversible changes in DNA do not accompany olfactory receptor gene choice.

Totipotency of neuronal nuclei

The two-step cloning procedure used to produce mice from neuronal nuclei generates mice in which the neuronal-derived ES cells give rise to all embryonic tissues while cells from the tetraploid host blastocyst contribute to the embryonic trophectoderm (Eggan et al 2001). Thus, cloning of lymphocytes or neurons via an ES cell intermediate did not reveal totipotency of a nucleus from a terminally-differentiated cell (Rossant 2002). To demonstrate totipotency of mature olfactory sensory neuron (OSN) nuclei, we transplanted nuclei from the cloned ES cells into enucleated oocytes to generate re-cloned mice (Eggan et al 2004). The cloned pups had enlarged placentas but displayed no overt anatomic or behavioural abnormalities and survived to become fertile adults, consistent with previous cloning experiments (Eggan et al 2001). These observations demonstrate that nuclei of terminally differentiated olfactory neurons can be reprogrammed to totipotency.

Development, cellular plasticity and the prospective for cell therapy

Immune rejection is a frequent complication of allogeneic organ transplantation due to immunological incompatibility. To treat the graft versus host disease (GVHD), immunosuppressive drugs are routinely given to transplant recipients, a treatment that has serious side effects. ES cells derived by nuclear transplantation are genetically identical to the patient's cells, thus eliminating the risk of immune rejection and the requirement for immunosuppression (Hochedlinger & Jaenisch 2003). Moreover, ES cells provide a renewable source of replacement tissue allowing for repeated therapy whenever needed. Indeed, a recent study demonstrated for the first time that NT could be combined with gene therapy to treat a genetic disorder (Rideout et al 2002). A key issue of transplantation medicine is the availability of isogenic functional cells that are of sufficiently high quality and can be obtained in large quantities. Therapeutic cloning would, in principle, solve this problem. Indeed, the recent generation of human embryonic stem cells by nuclear transfer into human eggs has provided evidence that this approach is a technically feasible strategy for the treatment of human disease. Yet, serious obstacles such as the availability of human eggs and ethical considerations impede the application of therapeutic cloning for the treatment of patients suffering from disorders such as Parkinson's disease or diabetes. As an alternative approach adult stem cells have been proposed because numerous studies have claimed that these cells can give rise to many or all cells of the adult by a process designated as 'transdifferentiation'. However, the process of transdifferentiation is remarkably inefficient, and some of the experimental claims have been difficult to reproduce or have alternative explanations (Alvarez-Dolado et al 2003, Terada et al 2002, Wang et al 2003, Ying et al 2002).

As depicted in Fig. 2, the nuclear transplantation from a differentiated donor cell into the egg can increase nuclear potency. This is accomplished by the action of reprogramming factors which are present in the egg's cytoplasm and which induce the genome to assume an epigenetic conformation that is appropriate for an embryonic state. It will be a major goal of future work to use the nuclear transfer technology as an experimental tool for defining the nature of the egg's reprogramming factors and the mechanisms of their action. Indeed, a recent report has demonstrated that *Oct4* is reactivated in mammalian somatic nuclei after having been transplanted into *Xenopus* oocytes (Byrne et al 2003). Thus, it may be possible to understand the molecular basis of epigenetic reprogramming and to establish rational strategies for altering the potential of somatic cells. The

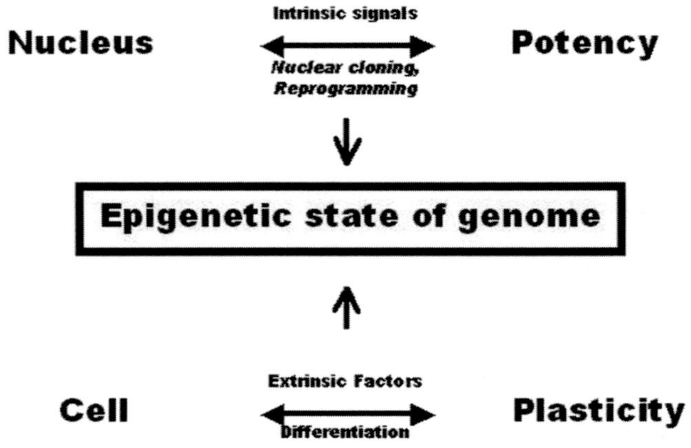

FIG. 2. Nuclear potency, cell plasticity and epigenetic conformation.

hope would be that such an approach eventually would allow the reprogramming of a patient's cells into different embryonic stem cells that could be used for cell replacement therapy without the need for nuclear transplantation.

Acknowledgements

The work summarized in this article has been supported by grants from the National Institutes of Health.

References

Alvarez-Dolado M, Pardal R, Garcia-Verdugo JM et al 2003 Fusion of bone-marrow-derived cells with Purkinje neurons, cardiomyocytes and hepatocytes. Nature 425:968–973

Betts D, Bordignon V, Hill J et al 2001 Reprogramming of telomerase activity and rebuilding of telomere length in cloned cattle. Proc Natl Acad Sci USA 98:1077–1082

Bird A 2002 DNA methylation patterns and epigenetic memory. Genes Dev 16:6–21

Boiani M, Eckardt S, Scholer HR, McLaughlin KJ 2002 Oct4 distribution and level in mouse clones: consequences for pluripotency. Genes Dev 16:1209–1219

Bortvin A, Eggan K, Skaletsky H et al 2003 Incomplete reactivation of Oct4-related genes in mouse embryos cloned from somatic nuclei. Development 130:1673–1680

Byrne JA, Simonsson S, Western PS, Gurdon JB 2003 Nuclei of adult mammalian somatic cells are directly reprogrammed to oct-4 stem cell gene expression by amphibian oocytes. Curr Biol 13:1206–1213

Chun J, Schatz DG 1999 Rearranging views on neurogenesis: neuronal death in the absence of DNA end-joining proteins. Neuron 22:7–10

Daniels R, Hall V, Trounson AO 2000 Analysis of gene transcription in bovine nuclear transfer embryos reconstructed with granulosa cell nuclei. Biol Reprod 63:1034–1040

Di Berardino MA 1980 Genetic stability and modulation of metazoan nuclei transplanted into eggs and oocytes. Differentiation 17:17–30

Di Berardino MA, McKinnell RG, Wolf DP 2003 The golden anniversary of cloning: a celebratory essay. Differentiation 71:398–401

Eggan K, Akutsu H, Hochedlinger K, Rideout W, Yanagimachi R, Jaenisch R 2000 X-chromosome inactivation in cloned mouse embryos. Science 290:1578–1581

Eggan K, Akutsu H, Loring Jet al 2001 Hybrid vigor, fetal overgrowth, and viability of mice derived by nuclear cloning and tetraploid embryo complementation. Proc Natl Acad Sci USA 98:6209–6214

Eggan K, Baldwin K, Tackett M et al 2004 Mice cloned from mature olfactory sensory neurons. Nature 428:42–49

Fulka J Jr, Miyashita N, Nagai T, Ogura A 2004 Do cloned mammals skip a reprogramming step? Nat Biotechnol 22:25–26

Gurdon JB 1999a Genetic reprogramming following nuclear transplantation in Amphibia. Semin Cell Dev Biol 10:239–243

Gurdon JB 1999b Genetic reprogramming following nuclear transplantation in Amphibia. Semin Cell Dev Biol 10:239–243

Hochedlinger K, Jaenisch R 2002a Nuclear transplantation: Lessons from frogs and mice. Curr Opin Cell Biol 14:741–748

Hochedlinger K, Jaenisch R 2002b Monoclonal mice generated by nuclear transfer from mature B and T donor cells. Nature 415:1035–1038

Hochedlinger K, Jaenisch R 2003 Nuclear transplantation, embryonic stem cells, and the potential for cell therapy. N Engl J Med 349:275–286

Humpherys D, Eggan K, Akutsu H et al 2002 Abnormal gene expression in cloned mice derived from ES cell and cumulus cell nuclei. Proc Natl Acad Sci USA 99:12889–12894

Jaenisch R 1997 DNA methylation and imprinting: why bother? Trends Genet 13:323–329

Jaenisch R 2004 The biology of nuclear cloning and the potential of embryonic stem cells for transplantation therapy. Appendix N in Monitoring stem cell research: a report of the President's Council of Bioethics. Washington DC, Government Printing Office, p 385–434

Jaenisch R, Bird A 2003 Epigenetic regulation of gene expression: how the genome integrates intrinsic and environmental signals. Nat Genet 33:245–254

Kang Y, Koo J, Park J et al 2001 Aberrant methylation of donor genome in cloned bovine embryos. Nat Genet 28:173–177

Lanza R, Cibelli J, Blackwell C et al 2000 Extension of cell life-span and telomere length in animals cloned from senescent somatic cells. Science 288:665–669

Lanza RP, Cibelli JB, Faber D et al 2001 Cloned cattle can be healthy and normal. Science 294:1893–1894

Liu L 2001 Cloning efficiency and differentiation. Nat Biotechnol 19:406

Nagy A, Rossant J, Nagy R, Abramow-Newerly W, Roder JC 1993 Derivation of completely cell culture-derived mice from early-passage embryonic stem cells. Proc Natl Acad Sci USA 90:8424–8428

Ogonuki N, Inoue K, Yamamoto Y et al 2002 Early death of mice cloned from somatic cells. Nat Genet 30:253–254

Rehen SK, McConnell MJ, Kaushal D, Kingsbury MA, Yang AH, Chun J 2001 Chromosomal variation in neurons of the developing and adult mammalian nervous system. Proc Natl Acad Sci USA 98:13361–13366

Reik W, Dean W, Walter J 2001 Epigenetic reprogramming in mammalian development. Science 293:1089–1093

Rideout WM 3rd, Eggan K, Jaenisch R 2001 Nuclear cloning and epigenetic reprogramming of the genome. Science 293:1093–1098

Rideout WM 3rd, Hochedlinger K, Kyba M, Daley GQ, Jaenisch R 2002 Correction of a genetic defect by nuclear transplantation and combined cell and gene therapy. Cell 109:17–27

Rossant J 2002 A monoclonal mouse? Nature 415:967–969

Tamashiro KL, Wakayama T, Akutsu H et al 2002 Cloned mice have an obese phenotype not transmitted to their offspring. Nat Med 8:262–267

Terada N, Hamazaki T, Oka M et al 2002 Bone marrow cells adopt the phenotype of other cells by spontaneous cell fusion. Nature 416:542–545

Tian XC, Xu J, Yang X 2000 Normal telomere lengths found in cloned cattle. Nat Genet 26:272–273

Vassar R, Ngai J, Axel R 1993 Spatial segregation of odorant receptor expression in the mammalian olfactory epithelium. Cell 74:309–318

Wakayama T, Yanagimachi R 1999 Cloning of male mice from adult tail-tip cells. Nat Genet 22:127–128

Wakayama T, Yanagimachi R 2001 Mouse cloning with nucleus donor cells of different age and type. Mol Reprod Dev 58:376–383

Wakayama T, Shinkai Y, Tamashiro KL et al 2000 Cloning of mice to six generations. Nature 407:318–319

Wang X, Willenbring H, Akkari Y et al 2003 Cell fusion is the principal source of bone-marrow-derived hepatocytes. Nature 422:897–901

Weissman IL, Anderson DJ, Gage F 2001 Stem and progenitor cells: origins, phenotypes, lineage commitments, and transdifferentiations. Annu Rev Cell Dev Biol 17:387–403

Wilmut I, Schnieke AE, McWhir J, Kind AJ, Campbell KH 1997 Viable offspring derived from fetal and adult mammalian cells. Nature 385:810–813

Yamazaki Y, Makino H, Hamaguchi-Hamada K et al 2001 Assessment of the developmental totipotency of neural cells in the cerebral cortex of mouse embryo by nuclear transfer. Proc Natl Acad Sci USA 98:14022–14026

Ying QL, Nichols J, Evans EP, Smith AG 2002 Changing potency by spontaneous fusion. Nature 416:545–548

Young LE, Sinclair KD, Wilmut I 1998 Large offspring syndrome in cattle and sheep. Rev Reprod 3:155–163

Zhang X, Firestein S 2002 The olfactory receptor gene superfamily of the mouse. Nat Neurosci 5:124–133

DISCUSSION

Gearhart: Your melanoma experiments remind me a bit of work many years ago by Gary Pierce. He would take different stage tumours and transfer these melanoma cells back into embryonic sites (Gerschenson et al 1986). He found that they normalized and integrated, and he didn't see tumours. These were selected cells, and once the cells got far enough they would never behave themselves. But this is an example of a cellular effect on nuclear factors.

Jaenisch: I think this fits well with the stem cell concept of cancer. We see this with the melanoma population which is a heterogeneous cell population. We got another ES line from this melanoma which didn't do anything and turned out to be tetraploid. Clearly, there are restrictions on what you can clone, and on the potential of cloned cells.

Weissman: In the teratocarcinoma, where you showed a high level of development of ES lines, you then showed that they don't all germline. Your final conclusion was that this was evidence of a genetic rather than epigenetic change. If you had taken the EC cells, made a teratocarcinoma and then from the teratocarcinoma from one of those neural cells from the line that made neural cells, you would expect to have the same low frequency as the cell that had gone through epigenetic changes to make a neural cell. Is that right?

Solter: It can't be the case. All his cells are aneuploid, and I don't think you can make a mouse from aneuploid cells.

Weissman: I am not asking him to make a mouse. All I am asking is whether he will lose the frequency that he had — the high level frequency of making an ES line to transfer if you allow them to differentiate. You could show that those cells do go through epigenetic changes and it alters their capability.

Jaenisch: The reason why there is such a high efficiency of nuclear transfer is because these cells all express these *Oct4*-like genes. We know that in every embryo and ES cell all of these 70 genes are expressed. But in the blastocysts derived from a fibroblast, or from another somatic cell, they are not. It is stochastic: some genes are expressed and some are not. There is some pressure in culture to activate these genes.

Goldman: With regards to the neuronal experiments you did, I was wondering about the choice of olfactory sensory epithelium. I can understand why you did this in terms of the olfactory receptor protein rearrangement, but there is a downside in that you cannot really claim that you're studying CNS or post-mitotic neurons. The olfactory sensory epithelium is not CNS; I think it's neural crest, so it is peripheral nervous system.

Jaenisch: No, I think it is CNS but I may be wrong.

Lovell-Badge: It is placode derived and therefore ectodermal in origin.

Goldman: So as a placodal mitotic sensory epithelium, it would be ectodermal in origin, but not neuroectodermal. It's really not CNS. So the neurons generated by the olfactory epithelium are post-mitotic, but the epithelium is rapidly turning over. Would you consider this representative of the CNS?

Jaenisch: We did the same with the CNS, with cortical neurons. The efficiency of making mice is as high, and the mice are normal. Certainly, cortical neurons can be put back into cycle. They were from postnatal mice and we labelled them with a CAM kinase Cre. A few years back there was a publication where they used neurons from fetal brains. They used neurons from areas of mature neurons and also from immature neurons. When they used immature neurons they got better development than from the mature neurons. They concluded there may be some genetic rearrangements which prevent neural cloning. There was no real evidence for this, apart from evidence suggesting that post-mitotic neurons have a lot of aneuploidy. 10% of them have lost the X or the Y. What they suggested was consistent with this type of model.

Solter: This is a frequent problem, and has been mentioned repeatedly. No matter how this experiment is designed, the tetraploid rescue doesn't demonstrate that the donor genome is capable of complete reprogramming, since it did not participate in the formation of extra-embryonic membranes. Did you try to clone by nuclear transfer the ES cells derived from blastocysts which themselves were derived from eggs containing the genome of neural cells? If this would work, it would demonstrate that the genome of the original neuron is capable of complete reprogramming.

Jaenisch: We did that experiment. It was after the B and T cell cloning when people said that it didn't show totipotency because it didn't make placenta. So we re-cloned by nuclear transfer ES cells cloned from the olfactory neurons, and we got mice. The nucleus of the ES cells is truly totipotent. It was important to do this experiment.

Gurdon: This would be a good time to ask what the difference is in the developmental potential of ES versus EG cells. Is there a real difference?

Jaenisch: Yes, I think there is a major difference. EG cells will never make mice because of the imprinting problem.

Gurdon: They can make a lot of different cell types. How many?

Gearhart: We don't get teratomas so we don't get an output where we can survey a lot.

Jaenisch: They make fine chimeras.

Lovell-Badge: They can contribute to the germline too.

Jaenisch: You can't do nuclear transfer because they are different from ES cells.

Grompe: I'd like to raise the question of the genetic stability of ES cells. I have worked on DNA repair before, and I'm surprised that ES cells can be as stable as they are through many passages. Is the fact that mice can come out of this because of a selection process in the blastocyst?

Jaenisch: We have looked at this carefully. In any good ES cell line, between 20 and 80% are aneuploid cells. The line with 80% aneuploid cells can make perfect tetraploids. Of course, the cells that contributed to the animal were all diploid. So I think that these cells are rather unstable, and that the aneuploid cells have a selective disadvantage. The ones that have a monosomy or trisomy grow slowly. It is almost an equilibrium between the aneuploid cells and the diploid cells. As long as you don't get chromosomal rearrangement that gives a selective advantage (trisomy 8 is one), you are fine. If we make a chimera we only select for the diploid cells. We have looked at what happens if we inject 15 cells into a blastocyst, and find that just one or two form the chimera—the rest are excluded. For tetraploids, if we inject 15 cells, only one or two donor cells make the mouse.

Grompe: Are the genetic rearrangements in the non-normal cells random, or are these rearrangements that give these cells a selective advantage in culture?

Jaenisch: We know that trisomy 8 comes up independently. We have seen it five or six times. If you have trisomy 8 these cells make the most beautiful colonies, they grow faster and they take over. When you do a knockout experiment, never take the beautiful colonies: take the crappy ones! This is a lesson we have learned.

Grompe: What about loss of heterozygosity (LOH) or mitotic recombination?

Jaenisch: We haven't looked carefully at this.

Gearhart: There is one report from Peter Stambrook's lab (Cervantes et al 2002) that looked at this in mouse ES cells, and the most common mutation is uniparental disomy, leading to LOH.

Evans: You can often drive it. If you have a knockout you can make it homozygous.

Jaenisch: If you fool around with methylation it goes up.

Grompe: This gets to Martin Evans' question: how do you quality control or validate human ES lines?

Lovell-Badge: What you are saying is right. You can get selection in culture for things that are abnormal. Often you will find over time that a line will get worse and worse at going germline. However, it is possible to subclone from the line and find cells that will work. There is a lot of variation in the population of ES cells. Going back to an earlier discussion we had, I think this reflects how much we still don't know about the best conditions in which to grow these cells. Martin mentioned work inhibiting the Wnt pathway, can we now maintain all the cells in a more uniform state?

Weissman: I'd warn you about the Wnt pathway. They have an inhibitor of GSK3β. This has many functions in many parts of the cell, so I would wait on that until it is independently replicated.

Lovell-Badge: I was just using this as an example. You can take any cell from the epiblast at 6 d and it will be able to do anything if it is moved around in the embryo. You should be able to get an ES cell from every cell in a 6 d embryo, although attempts to do so have failed.

References

Cervantes RB, Stringer JR, Shao C, Tischfield JA, Stambrook PJ 2002 Embryonic stem cells and somatic cells differ in mutation frequency and type. Proc Natl Acad Sci USA 99:3586–3890

Gerschenson M, Graves K, Carson SD, Wells RS, Pierce GB 1986 Regulation of melanoma by the embryonic skin. Proc Natl Acad Sci USA 83:7307–7310

General discussion II

Weissman: The fact that we have mouse strains with different abilities to make embryonic stem (ES) cells from blastocysts means that we can now do F2s and backcrosses to the mouse strains to try to map the genes involved. I know that you have done this, Rudi Jaenisch: what were the results?

Jaenisch: This is work we did a number of years ago, trying to answer the question of what really makes 129 mice so efficient at making ES cells while with other strains it is impossible. We identified mouse strains that we could not make ES cells from. We found a few, including FVB. We made a cross between 129 and FVB, and the F1s could make ES cells but only with about half the efficiency of the 129 parent. Then we made N1 and N2 crosses and the efficiency went down. It turned out that there were multiple genes involved so we gave up. However, this reinforces the idea that there clearly are genetic components to the ES cell-making ability.

Weissman: The problem with N2 is that in order to make a blastocyst to test whether you have a line, you still have to do the mating. It might have been better if they had already made recombinant inbred lines which you could go back and test. Certainly, there must be recombinant inbred lines.

Jaenisch: We tried some of those, but it became too much work.

Weissman: In recombinant inbred lines they do brother–sister mating after the first F2.

Jaenisch: I don't think they have done it with FVB.

Evans: I have tried FVB and I agree, I couldn't do it. My conclusion wasn't a genetic one, though. I thought it was because of timing: the embryos were developing considerably larger and faster.

Jaenisch: You said you could do it from morula.

Evans: What I was unable to do in those experiments was use delayed blastocysts, which I always find the best way. Did you try that?

Jaenisch: No, we did not.

Gearhart: Is there any association here with the propensity of 129 mice to teratomas?

Jaenisch: I don't think so. Other inbred strains which don't make these teratomas are perfectly able to generate ES cells.

Lovell-Badge: There is a locus on chromosome 8 which seems to be involved. Do we know what that is?

Jaenisch: No. The locus on chromosome 8 may have a function that is involved in proliferation, and is probably not important for establishment of ES cells. The trisomy gives ES cells a proliferative advantage.

Soria: I'd like to raise the issue of the differentiation capabilities of various mouse and human cells lines. It would be useful to have data on this. The access to human ES cell lines is limited, and it is difficult to compare the results. At least in the mouse, our group has observed that an R1 mouse ES cell line is better at generating ectoderm derivatives, whereas D3 mouse ES cell line may be better for endoderm. If this also happens with human ES cells, considering that the *in vitro* differentiation protocols are so long, it will take years to know which cell line is best for a particular differentiated cell.

Minger: One of the things we will do in the UK with most of the people who have access to cells and are making cell lines is to take all the cell lines that are currently in the UK and rigorously characterize them using a variety of standardized markers and techniques. The stem cell bank will make verified antibodies available to every one. We will look at all the usual antibodies, and there will be some microarray work. We will all use cell lines of the same passage on the same feeders and the same medium. We will also look at a lot of differentiation paradigms and do some epigenetics. Certainly, in the UK we have a concerted effort among 10 or 12 different labs that are all working with human ES cells.

Gearhart: There is also a voluntary international consortium.

Lovell-Badge: In the UK there is the National Institute of Biological Standards and Controls. This is where the bank will be.

Minger: Everyone who has a licence in the UK to generate cell lines is required to deposit their cells into the bank. Anyone can apply for access to these cells, and they can work out a material transfer agreement (MTA) with the originating institution.

Gearhart: Part of the issue here is that this will take time. We know it takes a great deal of effort even to maintain a few of these human lines in the lab.

Soria: I have a more specific question about the epithelial shape of cell lines from different species. Does this epithelial shape somehow restrict the differentiation potential? In the last paper we saw that epiblasts express a marker of the new ectoderm, but not of endoderm or mesoderm derivatives. Does this mean that these cell lines will be more restricted to ectoderm?

Evans: I don't think anyone knows. The slightly different growth characteristic actually reflects the embryology of the originating organism, whether it is human, mouse, pig or cattle. In humans, pigs and cows there is an embryonic disc on the surface of the slightly later blastocyst whereas in the mouse and rat we get an inner cell mass. The cells in the embryonic disc assume an epithelioid form but those in the mouse inner cell mass (ICM) do not until later when the embryonic ectoderm is formed?

Trounson: That's a hard question to answer. In a human, you might say that ES cells are totipotent because they can produce trophectoderm. You would think under these circumstances they might be closely related to the ICM. But human ES cells are very clearly different to mouse ES cells. I remember when I started working with human ES cells, of the colonies never looked like those of mouse ES cells. Mouse ES cells are responsive to leukaemia inhibitory factor (LIF), but the human cells are not responsive to LIF at all, and must have a different set of factors controlling their renewal and maintenance. They are made from ICMs that contain a layer of primitive endoderm. If you closely examine the cells individually in the human, they are quite variable in morphology. Using very specific markers we find quite a bit of variability within the colony.

Evans: I am not certain from my reading of the literature how many of these human cell lines have actually been rigorously cloned. Do you know?

Trounson: Very few. Perhaps one or two.

Evans: So you can't say that they are even pluripotent.

Trounson: No, not from a clonal point of view. They are extremely difficult to derive from single cells.

Minger: A Swedish group has cloned one of their lines and the efficiency was less than 0.1%.

Trounson: The problem is probably suboptimal culture conditions for clonal derivation of cell lines.

Evans: I would have thought that it doesn't matter if it is a low percentage. The critical question is whether these cells are multipotent.

Weissman: If you take a single marked cell from one line and grow it in another line, will it grow that way? You are letting the other line provide the factors.

Trounson: That's a good point and I don't know the answer.

Gearhart: Martin Evans, you have made the point that virtually none of the lines outside of mouse are ES cell lines.

Evans: It would be kinder to say that they haven't been characterized as well.

Gearhart: Should we change the name, or somehow indicate this difference? The reality is that we will never be able to meet your requirements for human ES cells.

Evans: You can get a lot further along than as of now.

Gearhart: We still don't have the gold standard.

Evans: I am much more worried about the sort of issue I have just raised: for very few of the lines that have been reported do we see multiple differentiation, either in tumours or *in vitro*. It is moreover important if we are to characterize these cultures as ES cell lines that the cells are themselves pluripotential and that the diverse differentiation is not the result of differentiation of a diverse set of precursor cells which have been co-cultured. The cloning experiments of mouse embryonal carcinoma (EC) and then ES cells were vital to their characterization. We expect a complex cell mix from an embryo to differentiate, but we need to know that it is coming from a single stem cell. I would like to see this pinned down.

Grompe: I agree with Irv Weissman. If you have poor cloning efficiency at low density you just use a low multiplicity of infection (MOI) gene transfer with an integrating vector, and pull out the green cells.

Gearhart: I am not questioning the issue of clonality. This has to be done, however, if you want to use chimera-making as a gold standard, we can't go that far with the human. But we want to get closer. I want to make a point of clarification on the term 'totipotency'. It is used differently in the USA and the UK, it seems.

Evans: I raised this issue because it has been used by the European parliament in trying to draft regulation. They will allow pluripotent cells to be handled but not totipotent cells. I think this is a complete misnomer.

Gearhart: In the USA, whoever wrote the initial NIH report in defining these terms gave totipotency two intertwined meanings. One is not only that it is a cell that can produce every other kind of cell but also it could organize it into an embryo. This was one of the problems when we were talking to our legislators. They envision if something is totipotent then you are growing embryos in a dish.

Weissman: Who wrote that?

Gearhart: This definition for totipotency appeared on the NIH Stem Cell information website for several years and was used, for example, by the then NIH Director, Harold Varmus, in a report to the US Senate, January 26, 1999.

Jaenisch: What is the issue of making chimeras with human stem cells? Is this something that is permissible?

Evans: I am not at all sure. It is so close to the knuckle that we'd have to take legal advice before we did it.

Jaenisch: It would be much more interesting to put a human ES cell in a gastrulating embryo, such as a chick embryo. Can we do this with human cells?

Evans: It is specifically forbidden to put human embryos, which we are allowed to do research on, into any animal.

Trounson: Human ES cell derivatives have been transplanted into the chick embryo to determine if they would form neurons.

Jaenisch: So is the chick acceptable but the mouse not?

Trounson: No, it's not unacceptable, but the issue is more one of what people will do when the paper comes out in a respectable journal, and then suddenly it is picked up by the media. You can imagine the headlines: 'Humice'. While this may not be biologically valid, it is important to be aware and sensitive to the interpretation.

Weissman: We are dealing with a group of people who say, 'Let's look at the extreme that could happen.' This is even true in the bioethics community. And even though the somatic nuclear transfer cells aren't embryos, the fact that there was a Dolly once means that people think of them as embryos. The community of bioethicists will take every extreme example. If one mouse embryo had a whole human brain and was born, they would find this difficult. The other issue is that your postdocs, graduate students and mouse caretakers will not necessarily wean animals at the time you want them to be weaned. If it has gone germline, even though the probability is infinitesimally low, a bioethicist will say that if there's any chance of all that there is a group of mice who are mating who have germline male and female gametes from human, then this is too risky. If you are going to do the experiment to help us understand these cells, you may have to do something to knock out particular developmental potentials.

Solter: One should first test whether any of this is feasible or true by using primate ES cells. Then one could see whether one could learn anything. What are you going to learn from the fact that some human ES cells can contribute to some mouse organs or tissues? We know that they can make all tissues.

Weissman: In 1998, the first thing you said to me is, 'Did you make chimeras with the human cells?' This was the only thing you wanted to know.

Solter: That was because at that stage we didn't know anything.

Weissman: The reason we'd like to do it, especially with genetically defined nuclear transfer of human pluripotent stem cell lines, is that we want them to develop in the organism. We want to see whether they will develop amyotrphic lateral sclerosis (ALS) and motorneuron degeneration if they came from a body cell of someone with ALS. There are many important questions we could try to answer that require the context of the developing body. Your proposal to put primate ES lines in mouse blastocysts is a good one.

Evans: We should remember that part of the problem with the outside world was one that we saw some years ago with putting human genes into mice.

Weissman: We got past that and it is now regulated not banned. We as a group are facing the heat already. If it is important to do this work, we should be as strong as the people who pushed the recombinant DNA into a regulatory rather than a banning environment.

Trounson: If we are having problems defining 'stemness', how are we going to define 'humanness' in that context? It is a bioethical issue that we could spend an enormous amount of time on. I would have thought that we are required to demonstrate function in these human cells in appropriate animal models anyway — possibly two animal models and probably including a primate. Sooner or later, if we want to have cell therapies, we will have to address this matter.

Wilmut: That would likely to be with a progenitor cell type, though, not a stem cell.

Gearhart: There are a lot of issues being discussed here but a major concern is whether there are appropriate animal models for these kinds of tests. Incidentally, we have just had a working group at Johns Hopkins and a report will be published later this year to discuss whether we should be working in non-human primates with human stem cells. The conclusion of this group was probably not. The issue that comes up is how good are animal models to demonstrate what we want. Are there good Parkinson's disease models that human cells could be placed in? I don't think the US Food and Drug Administration is going to object if we come in with the best data we have in an appropriate rodent model.

Weissman: We shouldn't allow anyone, unless it is for strict scientific reasons, to stop the march towards developing appropriate xenogeneic models: human cells in a mouse environment. If we give up that one step, then we say that research is open to political opinion.

Gearhart: The point I was making about this particular group is that they felt they could get this information in rodents without having to go to the non-human primate, which would raise a lot more issues.

Wilmut: We have concentrated on neurodegenerative diseases. What about other conditions? Are rodents always the best model? Would the same sensitivities be raised by the use of livestock for some conditions? One thing that we have wondered about is bone replacement, where rodents won't give the equivalent weight and pressure parameters to humans, but larger animals would.

Trounson: One of the factors that will prevent work with other species is the major species transplant barriers. If you are going to use human cells you will need to address immune rejection. If you remain in the mouse you can use an immunocompromised animal. When and if a tolerance is enabled, there may well be more reason to look at other appropriate tissue remodelling models, including rodents. With tolerance, our interests may be more about the nature and usefulness of the model.

Wilmut: It depends on exactly which system you are looking at, but you can create tolerance by putting cells into the fetus.

Jaenisch: I want to come back to the human–mouse chimeras. Injecting a human ES cell into a mouse blastocyst will be totally unsuccessful. We know that the human ES cells divide with a 40 h cell cycle and there is no way that they will incorporate into the mouse epiblast. It will be useful, however, to do this at a later stage. At E12–E14 human cells could incorporate into the mouse pancreas or brain, for example. I think this would be very useful. These would be a partial chimera: is that acceptable?

Weissman: It is possible, but it is a tougher injection.

Wilmut: Catherine Verfaillie, what MAPCs have you put into blastocysts?

Verfaillie: These are mouse and rat, and in the rat we haven't had anything chimeric, while in the mouse we have.

Trounson: Have you transplanted MAPCs across species?

Verfaillie: We have just done rat into mouse.

Weissman: If you put mouse ES cells into rat blastocysts, do they contribute? How big is the barrier in rats?

Evans: Probably the most informative experiments showing mouse ES cells forming chimeras in rats were performed by Phil Iannaccone and his colleagues (Iannaccone et al 1994, Brenin et al 1997). They had sought to isolate rat ES cells and showed that their cell lines made chimeras when injected into rat blastocysts which developed into adult rats. These showed a reasonable rate of chimerism in numbers of tissues. These experiments have to be reinterpreted as they later found that their cultures were predominantly mouse ES cells.

Gearhart: Martin Evans, as I recall your rat–mouse experiments, you had large clumps of rat cells and mouse cells. They never really co-mingled.

Evans: I suspect you are thinking of the results of Richard Gardner and Martin Johnson who were using rat ICMs in mouse blastocysts (Gardner & Johnson 1973).

Björklund: What would happen if you transferred a rat nucleus into a mouse blastocyst?

Jaenisch: It won't work. There is a paper where they put a human nucleus into a rabbit oocyte (Chen et al 2003). It was published after two years going round in circles, and it is plain wrong. It is not working because of the mitochondrial problem. You don't need respiration in cleavage, so they get interspecies blastocyts, but then the cells die. This is predicted for any interspecies fusion of cells. Human and chimpanzee fusion might be the exception because there is enough similarity between the nuclear genes and the mitochondrial genes.

Grompe: Is rat anti-mouse sufficiently close?

Jaenisch: No. They are pretty far away. This has been tried and it fails.

Solter: We did a nuclear transfer from one mouse species to another (Solter et al 1985). These embryos never developed, so even within the same genus different species might not work.

References

Brenin D, Look J, Bader M, Hiibner N, Levan G, Iannaccone P 1997 Rat embryonic stem cells: a progress report. Transplant Proc 29:1761–1765

Chen Y, He ZX, Liu A et al 2003 Embryonic stem cells generated by nuclear transfer of human somatic nuclei into rabbit oocytes. Cell Res 13:251–263

Gardner RL, Johnson MH 1973 Investigation of early mammalian development using interspecific chimaeras between rat and mouse. Nat New Biol 246:86–89

Iannaccone PM, Taborn GU, Garton RL, Caplice MD, Brenin DR 1994 Pluripotent embryonic stem cells from the rat are capable of producing chimeras. Dev Biol 163:288–292 (Erratum in Dev Biol 185:124–125)

Solter D, Aronson J, Gilbert SF, McGrath 1985 J Nuclear transfer in mouse embryos: activation of the embryonic genome. Cold Spring Harbor Symp Quant Biol 50:45–50

Nuclear reprogramming by *Xenopus* oocytes

J. B. Gurdon, J. A. Byrne and S. Simonsson

Wellcome Trust/Cancer Research UK Gurdon Institute, Tennis Court Road, Cambridge CB2 1QN, and Department of Zoology, University of Cambridge, UK

Abstract. The aim of this contribution is to relate amphibian nuclear transplantation work to prospects for stem cell creation and hence to the long-term aim of cell replacement in humans. The methods used include the transplantation of single somatic cell nuclei to enucleated unfertilized eggs of *Xenopus*, and also the transfer of multiple somatic cell nuclei to the nucleus (germinal vesicle) of a growing ovarian oocyte. A key difference between these types of recipient cell is that eggs immediately induce DNA replication in transplanted nuclei, whereas an oocyte induces no DNA replication, but directly reprograms an injected nucleus to a new pattern of transcriptional activity. We summarize the extent and success of past and current nuclear reprogramming in experiments with enucleated frog eggs, and also those carried out with growing oocytes. We discuss possible mechanisms of nuclear reprogramming, and the possible contribution of such knowledge for stem cell creation and cell replacement in humans.

2005 Stem cells: nuclear reprogramming and therapeutic applications. Wiley, Chichester (Novartis Foundation Symposium 265) p 129–141

The term 'nuclear reprogramming' describes situations in which a major change in gene expression takes place, so that the overall pattern of gene expression is switched from that characteristic of one cell type to that typical of another. The phenomenon is of interest because it throws light on the remarkable stability of the differentiated state of cells. The phenomenon is also important because it could provide an essential step in the creation of embryonic or stem cells from a differentiated cell and hence could contribute to the long-term aim of facilitating cell replacement in humans.

There are only two ways in which substantial nuclear reprogramming can be reliably achieved. One is by making heterokaryon cell hybrids. The other is by nuclear transplantation to eggs or oocytes. Good examples of nuclear reprogramming in hybrid cells are the following. If a chicken erythrocyte, a very small cell with very little cytoplasm, is fused to a large rapidly growing cultured cell, the erythrocyte nucleus enlarges to several times its volume and becomes very active in RNA synthesis, much like the recipient cultured cell (Harris 1970).

Another striking example is that of the fusion of a human amnion cell with a multinucleate chick myocyte (Blau et al 1983). In this case the hybrid heterokaryon starts to express human muscle proteins on its surface, showing that the amnion cell has been reprogrammed to express muscle genes characteristic of the large host muscle cell. In the case of nuclear transplantation to eggs or oocytes, a relatively tiny somatic cell is permeabilized and then injected into the cytoplasm of the egg. The resulting embryos generally show a pattern of embryonic gene expression, quite different from the range of genes expressed in the donor somatic cell (see Gurdon 1986 for review). With nuclear transfer to eggs, there is a huge excess of recipient cell cytoplasm over that of the donor cell. The purpose of this contribution is to summarize briefly the background to nuclear reprogramming in amphibian eggs and oocytes, to outline recent results, and to discuss possible mechanisms and the potential value of nuclear reprogramming.

Past work

Amphibian nuclear transfer experiments can be divided into two classes: (i) single somatic cell nuclei transplanted to enucleated eggs, resulting in embryo development; (ii) multiple somatic cell nuclei transplanted to non-enucleated growing oocytes (ovarian cells in first meiotic prophase). The main conclusions from these two classes of experiments are as follows.

As embryos develop, somatic cells become committed to particular pathways of differentiation. Cells are described as 'determined' when, by transplantation to heterologous sites, they cannot change from their committed pathway of differentiation. For example, from gastrulation onwards in the amphibian embryo, endoderm cells express endoderm-specific genes (Chalmers & Slack 1998, Zorn & Mason 2001) and these cells cannot be persuaded to follow other pathways of cell differentiation, even if transplanted to an environment of non-endodermal cells (Heasman et al 1984). The transplantation of nuclei from such cells to enucleated eggs shows that many such nuclear transplant embryos develop entirely normally, even to normal sexually mature adult frogs (Gurdon 1962a,b). Therefore, although endoderm cells themselves cannot change cell fate, the nuclei of these cells are completely reprogrammed back to an early embryo state so that they can participate normally in the whole range of developmental cell fates. When the nuclei of endoderm cells from neurula embryos are transplanted, about 10% of such nuclear transfers develop completely normally and are therefore described as totipotent (Gurdon 1960). Most such nuclear transplant embryos, however, do not develop normally and die at various developmental stages. But if we ask what proportion of determined endoderm cells contain nuclei able, after transplantation to an enucleated egg, to generate entirely unrelated cell types such as muscle and nerve, this proportion increases to about 20%. Comparable results are obtained when nuclei are transplanted to enucleated eggs from other cell types such as muscle cells, ectoderm cells, etc. The percentage of nuclear transplants that reach various stages of development is increased if the results of serial nuclear transplantation and of grafts from earlier nuclear transplant embryos are taken into consideration (Gurdon et al 2003). The figures quoted above include allowance for these experimental manoeuvres.

When nuclei are transplanted from cells of differentiated tissues, the frequency with which normal or partially normal development is obtained decreases. Nevertheless, a small percentage of fertile adult male and female frogs have been obtained from feeding tadpole intestinal epithelium and a higher proportion of such nuclear transplants develop to the muscular-response stage of tadpoles (Gurdon 1962b).

The results so far summarized depend on the morphological assessment of nuclear transplant embryo development. The use of molecular markers has been used to throw light on the mechanism of nuclear reprogramming. The markers used include general measures of nuclear activity such as the initiation of DNA synthesis and transcription of ribosomal RNA genes as well as individual genes characteristic of early development, often including those that encode transcription factors. Two conclusions can be drawn from the expression of molecular markers in nuclear transplant embryos. First, as soon as new zygotic transcription starts (mid-blastula), the genes expressed are those characteristic of normal embryos grown from fertilized eggs (Chan & Gurdon 1996). The timing of new gene expression is the same as in normal development. There are some quantitative irregularities affecting the amount of new gene expression, and these may help to account for some of the abnormalities often seen in nuclear transplant embryos (Byrne et al 2002). Second, we have seen, so far, no evidence of any residual gene expression characteristic of the donor cells. Therefore the reprogramming activity is remarkably effective and has already switched gene expression of transplanted nuclei back to an embryonic state within only a few hours of their transplantation to eggs.

The basic principles that emerge from amphibian nuclear transfer work are (i) the conservation of the genome during cell differentiation, and (ii) the reversibility of the epigenetic state of differentiated cells.

Recent work with oocytes

It is difficult to explore the mechanism of nuclear reprogramming using somatic cell nuclear transfer to eggs. This is because the immediate effect of egg cytoplasm on transplanted nuclei is to induce replication of DNA and chromosome assembly. Therefore events that immediately follow the transplantation of nuclei to eggs may be causally connected with DNA replication and chromosome assembly rather than with transcriptional reprogramming.

For this reason attention was paid some years ago to the use of oocytes as recipients for transplanted nuclei. Oocytes are growing egg cells in the ovary, in first meiotic prophase. These oocytes can not be fertilized and have no ability to respond to injected nuclei by undergoing either DNA replication or cell division. Somatic cell nuclei injected into oocytes undergo changes in gene expression as judged by the synthesis of new proteins. In cross species experiments, new proteins are detected that are encoded by donor nuclei and not by activation of new recipient oocyte genes (De Robertis & Gurdon 1977). Therefore it seems that oocytes can reprogram nuclei for new gene expression in the complete absence of DNA replication or cell division.

Recently we have turned our attention back to the use of oocytes to analyse the mechanism of nuclear reprogramming. In particular we have found that the nuclei of adult mouse thymocytes or human blood cells can be reprogrammed by injection into *Xenopus* oocytes (Byrne et al 2003). Evidence of reprogramming comes, in this case, from the induced expression of the stem cell marker gene *Oct4*. To be sure that the mammalian *Oct4* gene is induced, rather than expression of a host oocyte gene, we have been able to sequence the new transcripts. This makes it completely clear that oocytes containing mouse nuclei express mouse *Oct4*, those containing human nuclei express human oct4 and those not injected with nuclei do not express *Oct4* at all. Because of the large size and huge abundance of *Xenopus* oocytes this type of nuclear transfer experiment offers the prospect of analysing the mechanisms and of identifying molecules involved in nuclear reprogramming.

Mechanisms of nuclear reprogramming

It is convenient, at this stage, to consider that two steps may be needed to achieve nuclear reprogramming. The first is to derepress somatically switched-off genes and the second is to select the appropriate choice of genes for new expression. One or other, but not both, of these steps needs to be specific. For example, the first decondensation stage could be completely unspecific such that all genes are opened up and made available for transcription. In this case, the second step would be specific, depending on the availability of a selected set of transcription factors. Of all the genes available for transcription, only those for which the oocyte possesses appropriate transcription factors would be transcribed. The second model for reprogramming would suppose that the decondensation step is itself specific, so that only a subset of all genes are decondensed and made available for transcription. This would be followed by the unspecific transcription of all available genes by RNA polymerase. Oocytes are well known for the intense and widespread transcription of their genes, leading to the spectacular lampbrush chromosomes characteristic of this stage of development (Callan & Lloyd 1960).

The history of nuclear transplantation, summarized above, strongly favours a process in which there is a continuous cycling of information between nucleus and cytoplasm. When cells reach a differentiated state, we suppose that this state needs to be continually maintained by the transcription of regulatory genes sending messenger RNA to the cytoplasm which is translated into protein that moves back into the nucleus to maintain the expression of the same genes (Fig. 1A). This situation would change only when signal factors are received from outside the cell or when an asymmetric division redistributes the determinant proteins. According to this concept, cells would maintain their differentiated state or progress further down one pathway of differentiation and would never reverse their differentiated state under normal conditions. However if a nucleus is removed from its normal cytoplasmic environment and introduced into a huge pool of cytoplasm of another cell, that nucleus will be soon forced to adopt the pattern of gene expression characteristic of the host cell due to the overwhelming abundance of the host cell's regulatory factors. As mentioned above, cell fusion and nuclear transfer experiments achieve just this situation (Fig 1B,C). It is clear that nuclei from differentiated cells resist an immediate switch of expression; they put up a bit of a fight, but eventually the overwhelming abundance of cytoplasmic regulatory proteins overcomes their resistance and will force them into a new pattern of gene expression, so long as there is time for this to happen. Often in the case of nuclei transplanted to eggs there is not sufficient time and the reprogramming process is incomplete. In the case of oocytes, time is on the side of nuclear reprogramming, since no cell division takes place, and nuclear reprogramming is efficient.

The future analysis of the mechanisms and identification of nuclear reprogramming molecules through the use of oocytes is likely to follow two routes. One is to define changes in the epigenetic state of somatic nuclei; is DNA demethylated in the promoter region of those genes that are reactivated? Are histones modified? Do particular proteins have to be removed from somatic nuclei to allow their repressed genes to be reactivated? On the other side of the problem, we want to identify what molecules an oocyte provides to select genes for reactivation. These may well include a range of transcription factors, and these may be identifiable by analysing the relevant parts of the promoters and enhancers of reprogrammed genes.

Future prospects

If it should become possible to understand the mechanism of nuclear reprogramming in amphibian oocytes, and to identify the molecules involved,

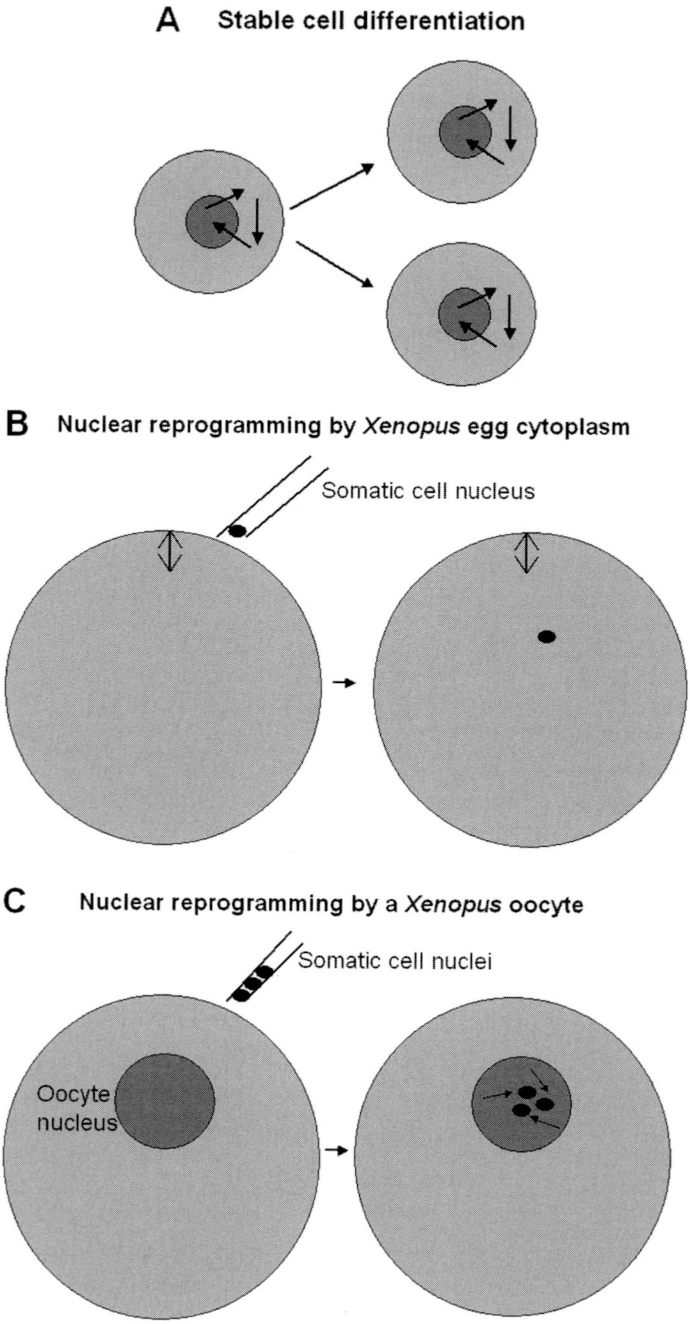
A Stable cell differentiation
B Nuclear reprogramming by *Xenopus* egg cytoplasm
Somatic cell nucleus
C Nuclear reprogramming by a *Xenopus* oocyte
Somatic cell nuclei
Oocyte nucleus

what further steps would be necessary to achieve the long-term goal of human cell replacement? At present we know only that *Oct4* is reprogrammed. Many other genes need to be activated in the right sequence and to the right extent, to create stem cells. If *Xenopus* oocytes reveal nuclear reprogramming mechanisms and molecules, we assume that homologous mammalian and human molecules can be identified by homology with *Xenopus* molecules. Eventually it may be possible, using reprogramming molecules, to directly convert adult mammalian blood cells to stem cells. Certainly there is a long way to go before human cell replacement can be achieved at least to some degree. But in view of the conservation of the genome during cell differentiation, and the reprogramming activity of eggs and oocytes, it seems inevitable that some kinds of cell replacement will be able in time to relieve some forms of human suffering.

Summary

It was established a long time ago that nuclei from differentiating cells can be rapidly reprogrammed by transplanting their nuclei into enucleated unfertilized eggs of *Xenopus*. Only after the blastula stage do embryos begin transcription. The pattern of gene transcription of such nuclear transplant embryos closely resembles that of normal embryos grown from fertilized eggs and not that of the donor cells. To investigate mechanisms of nuclear reprogramming, we transplant multiple somatic cell nuclei into growing oocytes, active in RNA but not DNA synthesis. We transplant about 100 nuclei from mammalian blood cells into a *Xenopus* oocyte and observe the activation of the stem cell marker gene *Oct4*. To explore the mechanisms of reprogramming we can inject a plasmid containing the mammalian *Oct4* promoter, and observe expression of this gene. By this approach, it is possible to begin to analyse the molecular mechanisms involved in reprogramming mammalian somatic cell nuclei to an embryonic pattern of gene expression.

FIG. 1. Diagrams to illustrate concepts of nuclear reprogramming. (A) Cells that have embarked on a particular differentiation pathway will continue to reprogramme their daughter cells for the same pattern of gene expression as themselves. Arrows indicate mRNA from nucleus (dark grey) to cytoplasm, protein synthesis in the cytoplasm (light grey), and passage of regulatory proteins from the cytoplasm back into nucleus. Therefore continuous reprogramming results in progression along the same differentiation pathway. (B) In somatic cell nuclear transfer to eggs, the amount of cytoplasm introduced with the transplanted nucleus is minimal by comparison with the huge volume of cytoplasm of the recipient egg. Consequently, the cyclic reprogramming of the somatic cell is overridden by regulatory molecules in the egg. (C) For oocyte nuclear transfers, multiple permeabilized somatic cells are injected into the nucleus of the oocyte (dark grey). Regulatory proteins within the oocyte nucleus enter the transplanted somatic nuclei and reprogramme them to oocyte and embryo gene expression.

Acknowledgements

This work was supported by grants from the BBSRC and the Wellcome Trust.

References

Blau HM, Chiu CP, Webster C 1983 Cytoplasmic activation of human nuclear genes in stable heterocaryons. Cell 32:1171–1180

Byrne JA, Simonsson S, Gurdon JB 2002 From intestine to muscle: nuclear reprogramming through defective cloned embryos. Proc Natl Acad Sci USA 99:6059–6063

Byrne JA, Simonsson S, Western PS, Gurdon JB 2003 Nuclei of adult mammalian somatic cells are directly reprogrammed to oct-4 stem cell gene expression by amphibian oocytes. Curr Biol 13:1206–1213

Callan HG, Lloyd L 1960 Lampbrush chromosomes of crested newts Triturus cristatus (Laurenti). Philos Trans R Soc Lond B Biol Sci 243:135–219

Chalmers AD, Slack JM 1998 Development of the gut in Xenopus laevis. Dev Dyn 212:509–521

Chan AP, Gurdon JB 1996 Nuclear transplantation from stably transfected cultured cells of Xenopus. Int J Dev Biol 40:441–451

De Robertis EM, Gurdon JB 1977 Gene activation in somatic nuclei after injection into amphibian oocytes. Proc Natl Acad Sci USA 74:2470–2474

Gurdon JB 1960 The developmental capacity of nuclei taken from differentiating endoderm cells of Xenopus laevis. J Embryol Exp Morphol 8:505–526

Gurdon JB 1962a Adult frogs derived from the nuclei of single somatic cells. Dev Biol 4:256–273

Gurdon JB 1962b The developmental capacity of nuclei taken from intestinal epithelium cells of feeding tadpoles. J Embryol Exp Morphol 10:622–640

Gurdon JB 1986 Nuclear transplantation in eggs and oocytes. J Cell Sci Suppl 4:287–318

Gurdon JB, Byrne JA, Simonsson S 2003 Nuclear reprogramming and stem cell creation. Proc Natl Acad Sci USA 100:11819–11822

Harris H 1970 Cell fusion. Clarendon Press, Oxford

Heasman J, Wylie CC, Hausen P, Smith JC 1984 Fates and states of determination of single vegetal pole blastomeres of X. laevis. Cell 37:185–194

Zorn AM, Mason J 2001 Gene expression in the embryonic Xenopus liver. Mech Dev 103: 153–157

DISCUSSION

Wilkins: The question of what molecules and processes carry out the methylation is an interesting one, but it seems to me that equally interesting is the identity of the molecule that directs the demethylase to its site of action. Are you doing footprinting experiments to identify those molecules that direct the demethylation?

Gurdon: Yes. I can't answer the question yet, but that is the way forward.

Jaenisch: You argued that the reprogramming depends on the cytoplasmic volume. If you use embryonic stem (ES) cells they have almost no cytoplasm, and they are dominant. If you fuse an ES cell with a thymocyte, they reprogramme the thymocyte nucleus and reactivate the X.

Gurdon: The ES cell has a much bigger cytoplasm.

Jaenisch: No, the ES cell has almost no cytoplasm, yet it reprogrammes very efficiently.

Gurdon: The ES cell volume must be hugely in excess of the volume of a thymocyte.

Jaenisch: An ES cell is certainly smaller than a fibroblast and possibly as small as a thymocyte.

Weissman: You say it's smaller than a thymocyte? There's nothing smaller than a thymocyte!

Gurdon: Let's just take the diameter of the cell and follow this argument through. A thymocyte is an extremely small cell. I can't believe that an ES cell is not bigger. It may have a big nucleus.

Jaenisch: It is smaller than a fibroblast.

Gurdon: It is a question of the ratio of the two volumes. What if the regulation molecules are in the nucleus?

Jaenisch: In this case, there must be something else in the ES cell that causes the reprogramming of the somatic cell other than just cytoplasm volume. I wanted to ask another question. You mentioned that there was a 2 d delay before activation was seen. When you inject an embryonic stem cell nucleus instead of a somatic cell nucleus into your oocytes, the stem cell nucleus is of course expressing *Oct4*. Would you now see the same delay as with the naked DNA?

Gurdon: We haven't done this. It would be a difficult experiment, because as you say the ES cell has already got *Oct4* transcripts in it. It would be an untidy experiment because you would have to look for an increase over a background, which would be difficult. I would add that if you take other cells, such as HeLa cells, they reprogramme quicker than adult thymocytes. I would assume that an embryonic cell would reprogramme quicker, too.

Evans: Could you use ES cell DNA and avoid these problems? The methylation should be different. My prediction would be as follows. If you use thymocyte nuclei it takes 2.5 days. If you use DNA it takes you 1 day. Perhaps if you used ES cell DNA it would take you 2–3 h.

Gurdon: Yes. That's possible. The objective of this experiment is to find a system where you can remove known components and relate them to the reactivation.

Evans: This is answering Rudi Jaenisch's question about whether an ES cell would work better.

Weissman: It is remarkable that you get these results with thymocytes. 70% of the thymocytes taken from a mouse thymus are destined to die in the thymus within 24 h. If you picked the small ones, virtually all will be destined to die. They are cells that did their DNA rearrangement to make the T cell receptor and it was the wrong T cell receptor for that mouse's thymus. They are therefore going down a default death pathway, and some will already be TUNEL positive. It is interesting that you can reactivate something that is so close to death. Perhaps in preparation for death the T cell is opening up DNA sites. The other interesting part of the genome that you are using is that you have these circles made of the intervening DNA of the β and α T cell receptor mutants which are closed together, and are now free from the chromosomal DNA. They usually represent V regions of particular genes. It would be interesting to know what chromatin state they are in. It is a trivial thing to find out.

Gurdon: That's an interesting point. I should add that we see *Oct4* reactivation with the nuclei of HeLa cells.

Solter: I assume that you are injecting nuclei.

Gurdon: No, not nuclei: they are permeabilized cells. This is one reason why any message which is still in the cell would confuse the analysis.

Solter: Can you be sure that when you inject thymus DNA versus thymus cells, that you inject DNA from the equivalent number of cells?

Gurdon: Yes.

Solter: I also have a developmental question. When you put the brain nucleus into the growing oocyte, can you remove the germinal vesicle and would the brain nucleus support oocyte maturation?

Gurdon: Almost certainly not. It is technically very hard to enucleate the oocyte. It can be done, but it is almost unreproducible. I can't imagine that they would substitute for the germinal vesicle.

Solter: When you showed the PCR results from successive days, my impression was that on day 4 there was something and day 5 was higher, and then day 9 was much lower. Does it really go down?

Gurdon: No, it is just variable. The problem is that by day 9 the oocytes are getting a bit worn out and fed up. Some keep going and some don't. This is particularly troublesome if they have had what we call a germinal vesicle injection. Survival is a great deal better if you haven't injected into the germinal vesicle. We aim to inject 100 nuclei, but it could be 50 or 150. We don't control this well so there is always a certain scatter.

Solter: Can you tell whether these nuclei divide once injected?

Gurdon: I know that they don't.

Grompe: I have a technical question. When you say you have injected DNA, do you mean stripped DNA, phenochloroform extracted with no chromatin on it?

Gurdon: Yes.

Grompe: So is the interpretation of your results that reprogramming activity does not operate at the level of chromatin?

Gurdon: My interpretation would be that it is a bit of both. The protein removal experiment seems to show that some derepression by the removal of donor nucleus protein helps, but it isn't the whole answer. I would imagine that you would have to remove the kinds of proteins that keep genes switched off in differentiated cells. You also have to do some demethylation of DNA and perhaps a lot of other things too.

Grompe: Does the DNA have to be rechromatinized?

Gurdon: The oocyte does that. It is a relatively quick process. The oocyte puts nucleosomes onto DNA with the correct spacing. In fact, if you give the oocyte chromatin with the wrong spacing, it then corrects it.

Grompe: Does this happen before you see the *Oct4* gene come on?

Gurdon: Yes. You can in principle have the removal of a protein or the modification of a protein remaining on the DNA, or rearrangement of the protein. That is when you put in nuclei or chromatin. DNA alone is assembled in the way the oocyte likes with its own components.

Nadal-Ginard: Is the difference between a HeLa cell and a thymocyte the difference between a cell that is cycling and one that isn't?

Gurdon: I would guess that a dividing cultured cell is generally more metabolically active and has more open genes. The thymocyte is particularly useful to us because it is heavily switched off and would take a more obvious re-programming to turn it back on again.

Wilmut: You said that the oocyte is transcriptionally very active. That is presumably for the specific functions of the oocyte. Do you know which other mammalian genes are switched on? Why is it that *Oct4* is switched on? Is there a comparable gene in amphibians?

Gurdon: There is no *Oct4* in amphibians, but there are three other *Oct* genes which are related but not the same. Regarding the other genes that are changed, that's a good question. Luckily we have a microarray expert in the next-door lab. We've done quite a few microarray experiments looking at what other genes change expression. The answer is at least another 50 or 100, some of which are ones you might predict and others of which you wouldn't. Some enzymes have no apparent role. In most cases we don't really know what genes are expressed in oocytes. *Oct4* is very convenient and works very well for us, so we are sticking with it.

Wilmut: Thinking in terms of using this sort of approach to pretreat nuclei before you then transfer them, would a nucleus that is beginning to transcribe as a mature oocyte nucleus be a good one to transfer to try to get subsequent development? I know that is not your objective.

Gurdon: What kind of nucleus?

Wilmut: It doesn't matter. You can choose what you like! Irv Weissman will give you a nice pure population of haematopoietic stem cells. You could put those in oocyte cytoplasm and that would alter gene expression. Would you anticipate that this would be advantageous if you could take the nuclei out or in some other way mimic this treatment, and then transfer them?

Gurdon: This is a complicated experiment. The reason for the complication is that all that manipulation is probably going to be fairly damaging. I anticipate a rather woolly kind of answer.

Jaenisch: Why did you inject cells and not nuclei?

Gurdon: The original logic is that we found when we purified nuclei they don't do very well. They are somewhat damaged.

Jaenisch: You don't care about damage: you just look for activation. Would it work with nuclei?

Gurdon: One could do this although I can't see the advantage. If we had a background of inherited message that was a nuisance there would be a reason for doing this. What would the question be that you are asking?

Jaenisch: I guess the system would be cleaner if you got rid of the cytoplasm.

Gurdon: The question you are getting at is does the cytoplasm in the permeabilized cell help or hinder or some way? This could be looked at. I don't see the system as particularly unclean at the moment, though.

Jaenisch: This whole demethylation during cleavage that is so well established in mouse has not been shown in *Xenopus*. *Xenopus* doesn't demethylate. But you find this now. That is interesting: there must be an active demeythlase involved because there is no DNA replication going on. This is one of the most mysterious activities in the methylation field. It would be very interesting if this activity is in the oocyte.

Gurdon: I agree.

Wilkins: It would also be of interest to know whether the gene reactivation is determined solely by local effects around the gene, or whether it is effected by chromosomal regional effects or nuclear compartment effects (if the gene is near a centromere or telomere, for example). Can you tell us where *Oct4* is on its particular mouse chromosome, and whether there are any translocation or inversion strains where the gene has moved?

Gurdon: I don't remember where it is.

Grompe: Why do you transfer the nucleus into the oocyte? Why don't you transplant cytoplasm, for example out of the oocyte into a differentiated cell?

Gurdon: You could take oocyte cytoplasm and put it into a differentiated cell, but you would have to remember what amount you would be adding. You would make something like a cytoblast, which was done in the old cell fusion days. This would be fused to another cell. These experiments have been done. Supposing you did this: what would you learn?

Grompe: It seems to me that in terms of the technical aspects, if there are cytoplasmic factors why couldn't one make a cytoplast out of ES cells and use it to reprogramme nuclei?

Sullivan: It is technically very difficult.

Jaenisch: Are these cytoplasmic factors or nuclear factors?

Gurdon: The factors are concentrated in the nucleus. What you would like to do is fractionate the cell and then you could find out which fraction is effective. We would need an assay for the fractions. It would be pretty horrendous experimentally to try to put these fractions into a cytoplast and fuse them into a cell.

Gearhart: Rudi Jaenisch is probably more interested than you in this because you don't want to produce animals, but is there any evidence that a mammalian nucleus put into these eggs for a period and then taken and put into a mouse egg has any greater developmental capability?

Gurdon: Ian Wilmut and I have talked about this experiment. It is technically difficult. The answer must be no at the moment. Our real aim is to find out what molecules and mechanisms are involved.

Gearhart: We want to jump ahead and see whether it will help a mammalian nucleus to be in that milieu.

Cardiac stem cells and myocardial regeneration

Bernardo Nadal-Ginard, Piero Anversa, Jan Kajstura and Annarosa Leri

Department of Medicine, Cardiovascular Research Institute, Vosburgh Pavilion, New York Medical College, Valhalla, NY 10595, USA

Abstract. Until recently, the accepted paradigm considered the adult mammalian heart a post-mitotic organ without intrinsic regenerative capacity where neither myocyte death nor new myocyte formation played any role in its homeostasis and could be safely ignored. We have recently identified in the adult mammalian myocardium a small cell population expressing surface antigens commonly associated with a variety of stem cells. These cells have the behaviour and potential of *bona fide* cardiac stem cells (CSCs): they are clonogenic, self-renewing and multipotent. Their presence has identified myocyte death and myocyte renewal as the two sides of the proverbial coin of cardiac homeostasis. Myocyte renewal depends on the differentiation of the CSCs into immature myocytes that divide two to four times before becoming terminally differentiated. Both *in vivo* and *in vitro* the progeny of a single CSC can generate the three major cell types of the myocardium: myocytes, smooth muscle and endothelial vascular cells. More interestingly, when directly injected or activated with growth factors in the post-ischaemic myocardium, these cells are able to reconstitute a functional ventricular wall. Thus, although in the adult heart most cardiac myocytes are permanently withdrawn from the cell cycle, the heart has an intrinsic regenerative potential and it is not a terminally differentiated organ.

2005 Stem cells: nuclear reprogramming and therapeutic applications. Wiley, Chichester (Novartis Foundation Symposium 265) p 142–157

Until the turn of this century, the accepted paradigm in cardiac biology and medicine considered the adult myocardium a post-mitotic organ without intrinsic regenerative capacity with a relatively constant but diminishing number of myocytes from shortly after birth to adulthood and senescence (Pasumarthi & Field 2002, Chien & Olson 2002, Rosenthal 2003). This belief has been based on the twin notions that:

- in the adult heart all myocytes, the cells that generate the contractile force responsible for the cardiac output, are terminally differentiated and, therefore, cannot be recalled into the cell cycle; and

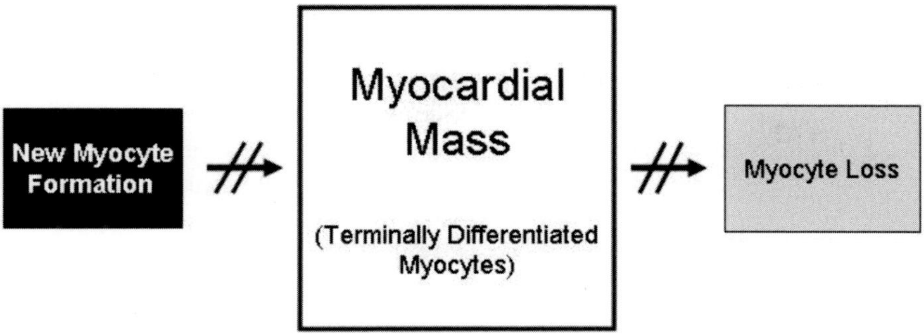

FIG. 1. Current view of cardiac homeostasis.

- that the myocardium, in contrast to other tissues such as bone marrow, liver, skin and even brain, lacks a stem cell population capable of generating new myocytes.

These two concepts have affected cardiovascular research and clinical cardiology in significant ways. If no new myocytes can be formed in adult life, it follows that all therapeutic interventions designed to preserve or improve cardiac function need to be oriented toward the preservation of the remaining myocytes. According to this point of view, myocyte death should normally be a very rare event if cardiac mass and an acceptable level of function have to be preserved throughout the lifespan of the individual. Moreover, by necessity, in a given heart all myocytes had to be as old as the individual. Thus, although often overlooked, according to the prevalent view, each and every myocyte of a 90-year-old person had to be at least 90 years old.

Despite several early observations documenting the existence of some cycling ventricular and atrial myocytes (Linzbach 1960, Rumyantsev 1964), the prevalent static view of the myocardium implied that both myocyte death and myocyte regeneration had little role in cardiac homeostasis. Although in recent years stem cells have been identified in and isolated from many adult tissues, including the blood, skin, liver, brain, skeletal muscle and intestinal tract among others (see Rosenthal 2003, Anderson et al 2001), the search for a cardiac stem cell had been considered futile because of the widely accepted lack of regeneration potential of this organ (Fig. 1).

Myocyte death and regeneration are central to cardiac cellular homeostasis

Evidence challenging the accepted wisdom has been slowly accumulating. First, it is obvious that if the heart lacks self-renewal properties, over time, even moderate rates of myocyte death would lead to the disappearance of the myocardial mass. Not surprisingly, therefore, the existence of myocyte death in the normal myocardium has been quite controversial (Schaper et al 1999, Anversa 2000, Kang & Izumo 2000). Nevertheless, there is now a consensus that myocyte death occurs in the normal and pathological heart of both human and experimental animals. Even the most conservative values indicate that, in the absence of new myocyte formation, the normal heart would lose most of its mass in a few decades and the senile and failing heart would disappear in a matter of several months or, at most, a few years (see Nadal-Ginard et al 2003a,b, Anversa & Nadal-Ginard 2002a). There is no doubt that myocyte death occurs throughout the lifespan of the organism, independently from cardiac disease. Therefore, the demonstration of cycling ventricular myocytes in the normal and pathological adult mammalian heart of several species, including humans was particularly intriguing (see Nadal-Ginard et al 2003a, Anversa & Nadal-Ginard 2002a for reviews).

In the normal heart, the rate of cell death increases with age but from middle-age on it is not balanced by an equivalent increase in myocyte formation. The excess in cell death over new myocyte formation results in a net reduction in myocyte number. This smaller number of myocytes hypertrophies to preserve the size of the myocardial mass and results in an old heart of normal or slightly decreased size but with enlarged parenchymal cells. Thus, myocyte death, hypertrophy and new myocyte formation characterize normal cardiac homeostasis and its evolution through time might explain why coronary artery disease and its complications are major risk factors in the elderly and why myocardial infarction is associated with increased morbidity and mortality in this population (Nadal-Ginard et al 2003a,b, Anversa & Nadal-Ginard 2002a).

Although the data summarized above provided an alternative view of adult cardiac homeostasis, it was not readily accepted because it lacked a proper biological context (Anversa & Kajstura 1998) and also failed to reconcile two apparently contradictory bodies of evidence: on one hand, the well-documented irreversible withdrawal of cardiac myocytes from the cell cycle soon after birth producing a heart constituted mainly by terminally differentiated myocytes (see Pasumarthi & Field 2002, Chien & Olson 2002) and, on the other hand, the presence of cycling myocytes undergoing mitosis and cytokinesis in the adult heart. Interestingly, the number of these cycling myocytes and, therefore, the rate of new myocyte formation, increases dramatically in old age (Chimenti et al 2003) and in response to acute (Beltrami et al 2001) and chronic overload (Urbanek et al 2003), suggesting that their generation responds to physiological demands. Together, these results raised questions about the nature of these cycling myocytes, their origin, as well as their physiological relevance.

It was soon noted that the cycling myocytes in the adult myocardium, as detected by expression of cell cycle related proteins and/or BrdU incorporation, were cells significantly smaller than the average myocyte (Nadal-Ginard et al 2003a, Anversa & Nadal-Ginard 2002a) and that many expressed functional telomerase activity (Leri et al 2001), an enzyme that so far has only been detected in cells that are either in or capable of entering the cell cycle. There was also an inverse correlation between DNA replication and expression of cell cycle inhibitors, such as P16, expressed in terminally-differentiated cells (Fig. 2). This phenotype raised the possibility that the cycling myocytes represented a subpopulation of newly born, immature cells, which had not yet irreversibly withdrawn from the cell cycle. This point of view was reinforced by the finding that in the acute phase of human myocardial infarction, the border zone and the myocardium remote from the damage exhibited very high levels of myocyte replication with newly formed myocytes of small size (Beltrami et al 2001). However, cases of sex mismatched cardiac transplants in humans, in which a female heart was transplanted into a male host (Quaini et al 2002), have shed light on the origin of these cycling myocytes and paved the way for the identification of cardiac stem cells.

The heart harbours stem-like cells which are responsible for myocyte regeneration

Previous work from our group had shown that bone marrow cells (BMCs) injected directly into ischaemic myocardium (Orlic et al 2001a) or mobilized into the peripheral circulation by systemic administration of cytokines (Orlic et al 2001b) were able to home to acutely infarcted myocardium, undergo significant expansion, differentiate into the three main cardiac cell types — myocytes, smooth and endothelial vascular cells — and organize into an anatomical and functional new myocardium. The cytokine regime used for BMC mobilization produces high levels of circulating bone marrow-derived multipotent cells. This BMC mobilization is just an amplification of a naturally occurring phenomenon because under normal conditions there are bone marrow-derived stem cells in the circulation (Orlic et al 2001b). These results, which have created significant controversy (Balsam et al 2004, Murry et al 2004, Chien 2004), raised the possibility that circulating cells colonize the myocardium and contribute to myocyte renewal. The cases of sex mismatched cardiac transplants in humans, where a female heart is transplanted into a male host, offered an ideal setting in which to test this hypothesis. The colonization and differentiation of host cells homed to the transplanted female heart can be identified by the presence of a Y chromosome in the host-derived cells. Shortly after transplantation the female hearts in the male hosts had a significant number of Y positive myocytes and coronary vessels (Quani et al 2002). Although there are some discrepancies among research groups about the degree of cardiac chimerism, most likely due to technical differences (Anversa & Nadal-Ginard 2002b), there is no argument that these male cells in the female hearts originated from host cells that colonized the heart after the transplant and subsequently differentiated. These cells are an irrefutable proof of new myocyte formation in the adult heart and presuppose the existence of mobile stem-like cells able to differentiate into the three main cardiac cell types: myocytes, smooth muscle and endothelial vascular cells. We hypothesized that these migrating myocyte-generating cells were likely to have many of the characteristics of stem or precursor cells, which should make them identifiable in the transplanted heart myocardium.

Cells of donor and recipient origin that express stem cell-related surface antigens—c-kit, Sca-1, and MDR1—but are negative for markers of the blood cell lineages (lineage negative, Lin^{neg}) were identified in each of the transplanted donor hearts. Most importantly, identical cells were found in the control hearts. Additionally, cardiac progenitor and precursor cells of male and female origin were detected by co-expression of stem cell antigens and transcription factors of the cardiocyte lineage or markers characteristic of endothelial and smooth muscle cells (Quaini et al 2002). Taken together, these observations indicated the existence of stem-like cells, which have the potential to regenerate the main components of the myocardium: contractile cardiocytes and coronary vasculature. The data, however, did not offer information on whether all cell types originated from one or several different precursors.

The primitive cells in the adult myocardium are true cardiac stem cells

To determine whether the putative cardiac stem cells detected in the human heart transplants and normal control hearts were true stem cells with cardiogenic potential, we isolated them from the hearts of experimental animals and tested their differentiation potential *in vitro* and *in vivo*. Undifferentiated cardiac cells expressing the stem cell market c-kit, the receptor for SCF (stem cell factor) and negative for all the blood lineage markers (Lin^{neg}) were isolated from the ventricles of adult rats and cloned. These cells have continued to proliferate more than 48 months after their original isolation without reaching growth arrest or senescence and >90% of cells maintain a diploid chromosomal complement. When cultured in bacteriological plates, these cells do not attach and form clones in suspension analogous to the pseudoembryoid bodies, neurospheres, formed by the neural stem cells (Reynolds & Weiss 1992, Gritti et al 1996). By analogy we have termed these clones ‘cardiospheres’. Each cardiosphere represents the progeny of a single cell. When allowed to differentiate, each cardiosphere produced three different cell types, each expressing biochemical markers characteristic of myocytes, smooth muscle, or endothelial cells. However, the morphological and functional differentiation of these cells was incomplete. At no time did we detect spontaneous contractions of the cells expressing myocyte markers. Conversely, skeletal muscle, blood cell lineages, and neural markers were not detected. Similar results were obtained when attached clones were allowed to form.

Despite the aborted differentiated phenotype, the continued growth of these cells over extended periods of time and the very high frequency of differentiation in the three main myocardial cell types, is a strong indication of their self-renewal potential. The fact that the progeny of a single clone gives origin to cells with biochemical markers of the myocyte, smooth muscle and endothelial cell lineages demonstrates that these cells are multipotent. Taken together, these results indicate that the Lin^{neg} c-kit^{pos} cells isolated from the adult myocardium have the *in vitro* properties expected from cardiac stem cells (CSCs): they are self-renewing, clonogenic and multipotent (Beltrami et al 2003). Similar results have been obtained with the cardiac Sca-1- and MDR1-positive cells. Cells isolated from mice, rats, dogs and human have very similar phenotypes. Other authors have also identified similar cells (Oh et al 2003, Matsuura et al 2004).

Cardiac stem cells regenerate a functional myocardium

The abortive differentiation of the CSCs when grown *in vitro* suggested that the culture medium was deficient in factors required for complete differentiation. To determine their ability to fully differentiate and reconstitute the myocardium when injected into syngeneic rats acutely after myocardial infarction, we injected 1×10^5 cells derived from a single clone and marked either with BrdU or by expression of the transgene enhanced green fluorescent protein (EGFP) into the myocardium. Cells were injected at two sites of the border of a 5-hour-old infarct in two groups of animals. Infarcted animals injected with equal volumes of PBS and sham-operated animals injected with the same number of cells were used as controls.

A band of BrdU-labelled or EGFP-positive regenerating myocardium was identified in $\sim$80% of the cell-treated infarcted hearts that was absent from all the controls. The regenerated myocardium had reduced the infarct size by $\sim$25% and was constituted by labelled small myocytes, capillaries and arterioles that matured over time. The new myocytes expressed cardiac sarcomeric proteins, N-cadherin and connexin 43 (main components of the zone adherens and gap junctions that allow mechanical and electrical coupling, respectively). The injected cells underwent an impressive degree of amplification. At 10 days after injection, the regenerating band contained $\sim14\times10^6$ myocytes with an average cell volume of $1500\,\mu m^3$, while at 20 days the band had $\sim13\times10^6$ larger myocytes with an average volume of $3400\,\mu m^3$, as compared to the average

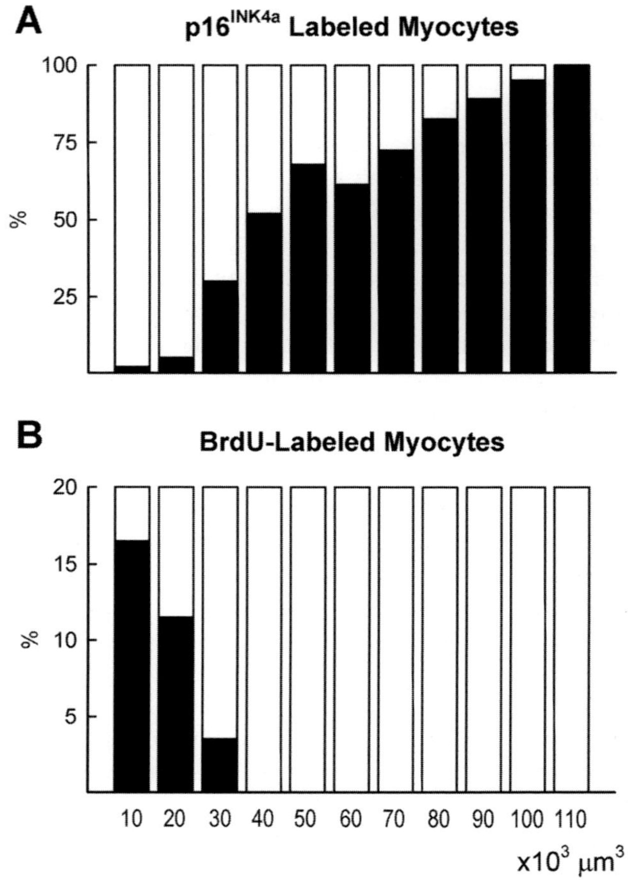

FIG. 2. Replicating myocytes are small.

volume of 20 000–25 000 μm^3 for adult rat myocytes. In addition, arteriolar density in the regenerating myocardium had reached mature values (55–60/mm^2) in these animals, while capillary density (1130/mm^2) was more comparable to the neonatal than to the adult heart (4000/mm^2), even at the later time point. All these characteristics reinforce the notion that at 20 days, the regenerated myocardium, constituted by new blood carrying vessels and myocytes, resembles the neonatal heart.

The regenerating myocardial cells positively affected cardiac performance and contraction reappeared in the infarcted ventricular wall. Cell implantation reduced infarct size and cavitary dilation; wall thickness, ejection fraction, + and −dP/dt significantly improved at 20 days. End-diastolic pressure and diastolic stress was 52% lower in treated rats. Thus, at 20 days the treated animals had a decreased diastolic load and improved overall pump function (Beltrami et al 2003).

The improved functional performance of the post-infarcted hearts suggests that the newly generated myocytes are functional and contribute to the overall contractility of the treated hearts. This point of view is further reinforced by the appearance of synchronous motility of the treated ventricular wall, as detected by echocardiography and the contractility parameters of isolated cells. Despite the dramatic size difference between new and surviving myocytes (new myocytes are ~1/10 the size of the spared ones), the contractility parameters of the new cells exhibit similar time to peak shortening, but with higher values of peak and velocity of shortening than spared myocytes. Therefore, from a structural and functional point of view, *in vivo* the cardiac stem cells generate *bona fide* myocardial tissue composed of cardiac myocytes and vasculature (Beltrami et al 2003).

The injected cells exhibited a tropism for the ischaemic tissue because all the regenerating cells were located in the necrotic area and not in the spared myocardium, despite the fact that the cells had been injected in the border zone between ischaemic and healthy myocardium. The stimulating role of ischaemia is further highlighted by the rapid disappearance of the transplanted cells when injected into the healthy myocardium of the mock-infarcted animals.

As was the case of the myocardial regeneration produced by the bone marrow-derived cells, there was no evidence of any significant cell fusion between the donor and host cells using the same criteria outlined in Nadal-Ginard et al (2003a). Whether or not the CSCs fulfil all the criteria that some believe are required before the label of 'stemness' can be applied to a given cell type (Anderson et al 2001, Wagers & Weissman 2004) is beside the point. Independently of what is the appropriate name for these cells, the evidence is convincing that the progeny of a single cell is able to produce the three main cardiac cell types and organize them into a functional myocardium (Beltrami et al 2003).

The presence of cells with CSC properties in the adult myocardium conclusively dispels the notion that the heart is a terminally differentiated organ without self-renewal potential and points toward this endogenous renewal potential as a tool for the treatment of heart diseases with a significant loss of myocytes. Because these CSCs regenerate both myocytes and small vessels, they might provide the new myocardium with self-renewing capabilities and might be attractive candidates for clinical myocardial regeneration.

The CSCs are distributed throughout the myocardium and can be stimulated *in situ* to regenerate lost myocardium

The demonstration of a cohort of CSCs in the myocardium raised the question of their location and distribution in the myocardium. CSCs are distributed unevenly throughout the heart. Although the majority are in the body of the left ventricle, their concentration per unit volume is significantly higher in the atria and apex than at the base and mid-region of the ventricle (Fig. 2). The atria and the apex are the regions of the heart that are exposed to lower levels of wall stress. On average there is one CSC per ~10 000 myocytes in a healthy myocardium or one CSC per every ~30 000 to 40 000 myocardial cells (does not include the cells with markers for progenitors and precursors, see below). This myocardial concentration of CSCs approximates that of HSCs in the bone marrow (Spangrude et al 2003) and is very similar in different species, including humans.

Hepatocyte growth factor (HGF) and insulin-like growth factor 1 (IGF-1) are produced by ischaemic myocardium (Ono et al 1997) and also synthesized by the CSCs, which also possess receptors for these cytokines (Torella et al 2005). *In vitro* and *in vivo* studies have documented that HGF is a chemoattractant for the CSCs (Torella et al 2005). IGF1 promotes cell survival, proliferation and differentiation. In addition, the activation of c-Met and IGF receptors on these primitive cells results in the induction of a characteristic set of integrins that might be responsible for the mobilization and growth of CSCs. The induction of the integrins $\alpha_V\beta_3$ and $\alpha_6\beta_4$ in the early stages of myocardial damage might facilitate the interaction with fibrinogen/fibrin and plasma fibronectin as well as mobilization towards the areas of damage. The late activation of $\alpha_4\beta_1$, $\alpha_2\beta_1$, and $\alpha_6\beta_4$ integrin receptors favours interaction with collagen and fibronectin during the organization of the scar. Thus, it was of interest to test whether local intra-myocardial administration of HGF and IGF-1 shortly after coronary artery ligation would result in the accumulation and activation of these cells in the infarcted area. As discussed in more detail elsewhere (Torella et al 2005), the results of local cytokine administration are indistinguishable from those obtained after the transplantation of the *in vitro* expanded CSCs. These results, together with the isolation and *in vitro* expansion of the CSCs, have obvious potential for clinical applicability.

Loss and/or ageing of CSCs results in a decrease of myocyte number and impaired ventricular performance

If cardiac cell homeostasis is dependent on myocyte and vascular cell regeneration from the CSCs, it can be predicted that loss of CSC function, either as a consequence of death or because they have become non-productive, should result in progressive myocyte loss and impaired ventricular function. This is exactly what happens in experimental animals and humans (Chimenti et al 2003, Torella et al 2004, Leri et al 2003).

Proliferation of mammalian cells, including human, is dependent on having functional telomeres. Telomere shortening in the absence of telomerase activity is one of the major causes of loss of proliferation in mammalian cells (Collins & Mitchell 2002, Harley et al 1990). Telomere attrition occurs with age and is proposed to the causal for replicative cell ageing. To test the effect of this type of ageing in the heart, we analysed the phenotype of telomerase knock-out mice, *Terc*$^{-/-}$. Progressive loss of telomere sequences in these animals eventually leads to loss of organismal viability after 3–6 generations (Lee et al 1998). Telomerase shortening at the second and fifth generation of *Terc*$^{-/-}$ mice was coupled with a profound attenuation in new myocyte formation, increased apoptosis and hypertrophy of the remaining myocytes. These cellular changes were concomitant with ventricular dilatation, wall thinning and ventricular dysfunction resulting in de-compensated eccentric hypertrophy and heart failure (Leri et al 2003). These results were confirmed in two additional animal models.

Cardiac stem cells play a fundamental role in normal cardiac cellular homeostasis

Despite the increasing body of data documenting the formation of new cardiac myocytes in the normal and diseased adult mammalian heart, the idea that normal and pathological cardiac homeostasis represents a balance between myocyte death and renewal has been met with significant resistance by the cardiovascular community. At least in part this resistance has been due to the fact that until now there has been a lack of biological understanding about the origin and fate of the cycling myocytes detected in the normal and pathological heart. The existence in the myocardium of stem cells capable of committing to the myocyte lineage provides an explanation and a biologically satisfactory context for the existence of cycling myocytes in the adult myocardium. The finding that some of these primitive cells *in situ* have already activated the myogenic programme, their ability to generate a progeny of millions of myocytes from a single cell, together with the small size of the cycling myocytes in the normal, pathological and regenerating heart indicates that these cycling myocytes are the progeny of the cardiac stem cells that have not yet reached a mature and terminally differentiated state. Unpublished data indicate that, once committed to the myocyte lineage, the progeny of a stem cell can undergo only three or four rounds of cell replication before withdrawing permanently from the cell cycle. Taken together, all these data highlight a mechanism for the continuous renewal of the myocardium—both myocytes and coronary vessels—throughout the lifespan of the individual. The CSCs are the source of new myocytes and vascular cells that throughout adult life regenerate those lost by the normal wear and tear. These CSCs go on overdrive in pathological conditions characterized by significant myocyte loss. Additionally,

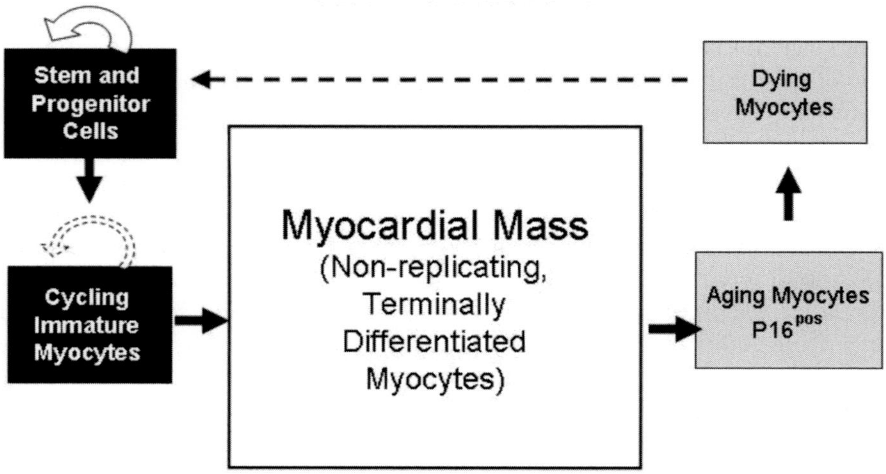

FIG. 3. Proposed view of adult cardiac homeostasis.

they provide a basis for the increase in myocardial contractile cells (hyperplasia) that complements the hypertrophic response when physiological and/or pathological demands impose an increased workload on the heart (Schaper et al 1990, Urbanek et al 2003) (Fig. 3). Thus, the adult heart, which like the brain is composed of mainly terminally differentiated parenchymal cells that do not re-enter the cell cycle, is not a terminally differentiated organ because it also contains stem cells that support its regeneration.

Prospects for inducing clinically relevant myocardial regeneration in acute or chronic ischaemic heart disease

Over the past two decades there have been significant advances in the management of acute and chronic myocardial infarction. Despite this progress, there remains a very large population of post-infarct survivors who will develop cardiac failure, leading to early death. Cardiac transplant remains the only therapeutic option available to many of these patients. Because of the ingrained concept of the myocardium as a terminally differentiated organ, it has been assumed that any attempt to replace the lost myocytes using cellular therapy would require the introduction of exogenous cells with myogenic potential into the myocardium. Recently, there has been a flurry of clinical trials for human myocardial therapy using bone marrow-derived cells and skeletal myoblasts (see Rosenthal 2003). The demonstration that the heart harbours stem cells capable of regenerating large amounts of functional myocardium (Beltrami et al 2003) explains our earlier observation of a robust regeneration in response to work overload (Urbanek et al 2003) and in the acute post-infarct (Beltrami et al 2001) but raises the question of why this regenerative response stops before the completion of the repair. The identification of the stem cells (Beltrami et al 2003) offers an opportunity to address this clinically important question and opens the tantalizing possibility that in the near future we might manipulate human CSCs *in situ* to home to the areas of damaged myocardium, proliferate and differentiate and, therefore, promote a functional cardiac repair without the need of introducing exogenous cells, as has been shown feasible in rodents (Torella et al 2004). Elucidation of the regulation of new myocyte formation will provide not only a better understanding of normal and pathological cardiac homeostasis but will open new avenues for therapeutic intervention. The extraordinary clinical need for some form of myocardial functional repair makes the understanding of the biological properties of these cells, and the harnessing of their regenerative capacity an exciting possibility that might help to maintain and restore cardiac function and, perhaps, reduce the need for cardiac transplantation.

References

Anderson DJ, Gage FH, Weissman IL 2001 Can stem cells cross lineage boundaries? Nat Med 7:393–395

Anversa P 2000 Myocyte death in the pathological heart. Circ Res 86:121–124

Anversa P, Kajstura J 1998 Ventricular myocytes are not terminally differentiated in the adult mammalian heart. Circ Res 83:1–14

Anversa P, Nadal-Ginard B 2002a Myocyte renewal and ventricular remodeling. Nature 415:240–243

Anversa P, Nadal-Ginard B 2002b Cardiac chimerism: methods matter. Circulation 106: e129–e131

Balsam LB, Wagers AJ, Christensen JL, Kofidis T, Weissman IL, Robbins RC 2004 Hematopoietic stem cells adopt mature hematopoietic fates in ischemic myocardium. Nature 428:668–673

Beltrami AP, Urbanek K, Kajstura J et al 2001 Evidence that human cardiocytes divide after myocardial infarction. N Engl J Med 344:1750–1757

Beltrami AP, Barlucchi L, Torella D et al 2003 Adult cardiac stem cells are multipotent and support Myocardial regeneration. Cell 114:763–776

Chien KR 2004 Stem cells: lost in translation. Nature 428:607–608

Chien KR, Olson EN 2002 Converging pathways and principles in heart development and disease. Cell 110:153–162

Chimenti C, Kajstura J, Torella D et al 2003 Senescence and death of primitive cells and myocytes lead to premature cardiac aging and heart failure. Circ Res 93:604–613

Collins K, Mitchell JR 2002 Telomerase in the human organism. Oncogene 21:564–579

Gritti A, Parati EA, Cova L et al 1996 Multipotential stem cellsfrom adult mouse brain proliferate and self-renew in response to basic fibroblast growth factor. J Neurosci 16: 1091–1100

Harley CB, Futcher AB, Greider CW 1990 Telomeres shorten during aging of human fibroblasts. Nature 345:458–460

Kang PM, Izumo S 2000 Apoptosis and heart failure: a critical review of the literature. Circ Res 86:1107–1113
Lee HW, Blasco MA, Gottlieb GJ, Horner JW, Greider CW, DePinho RA 1998 Essential role of mouse telomerase in highly proliferative organs. Science 392:569–574
Leri A, Barlucchi L, Limana F et al 2001 Telomerase expression and activity are coupled with myocyte proliferation and preservation of telomeric length in the failing heart. Proc Natl Acad Sci USA 98:8626–8631
Leri A, Franco S, Zacheo A et al 2003 Ablation of telomerase and telomere loss leads to cardiac dilatation and heart failure associated with p53 upregulation. EMBO J 22:131–139
Linzbach AJ 1960 Heart failure from the point of view of quantitative anatomy. Am J Cardiol 5:370–382
Matsuura K, Nagai T, Nishigaki N et al 2004 Adult cardiac Sca-1-positive cells differentiate into beating cardiomyocytes. J Biol Chem 279:11384–11391
Murry CE, Soonpaa MH, Reinecke H et al 2004 Hematopoietic stem cells do not transdifferentiate into cardiac myocytes in myocardial infarcts. Nature 428:664–668
Nadal-Ginard B, Kajstura J, Leri A, Anversa P 2003a Myocyte death, growth, growth and regeneration in cardiac hypertrophy and failure. Circ Res 92:139–150
Nadal-Ginard B, Kajstura J, Anversa P, Leri A 2003b A matter of life and death: cardiac myocyte apoptosis and regeneration. J Clin Invest 111:1457–1459
Oh H, Bradfute SB, Gallardo TD et al 2003 Cardiac progenitor cells from adult myocardium: homing, differentiation, and fusion after infarction. Proc Natl Acad Sci USA 100:12313–12318
Ono K, Matsumori A, Shioi T, Furukawa Y, Sasayama S 1997 Enhanced expression of HGF/c-Met by myocardial ischemia and reperfusion in a rat model. Circulation 95:2552–2558
Orlic D, Kajstura J, Chimenti S et al 2001a Bone marrow cells regenerate infarcted myocardium. Nature 410:701–705
Orlic D, Kajstura J, Chimenti S et al 2001b Mobilized bone marrow cells repair the infarcted heart, improving function and survival. Proc Natl Acad Sci USA 98:10344–10349
Pasumarthi KB, Field LJ 2002 Cardiomyocyte cell cycle regulation. Circ Res 90:1004–1054
Quaini F, Urbanek K, Beltrami AP et al 2002 Chimerism of the transplanted heart. N Engl J Med 346:5–15
Reynolds BA, Weiss S 1992 Generation of neurons and astrocytes from isolated cells of the adult mammalian central nervous system. Science 255:1707–1710
Rosenthal N 2003 Prometheus's vulture and the stem cell promise. N Engl J Med 346:5–15
Rumyantsev PP 1964 DNA synthesis and nuclear division in embryonical and postnatal histogenesis of myocardium. Arch Anat 47:59–65
Schaper J, Elsasser A, Kostin S 1999 The role of cell death in heart failure. Circ Res 85:867–869
Spangrude GJ, Perry SS, Slayton WB 2003 Early stages of hematopoietic differentiation. Ann NY Acad Sci 996:186–194
Torella D, Rota M, Nurzynska D et al 2004 Cardiac stem cell and myocyte aging, heart failure, and insulin-like growth factor-1 overexpression. Circ Res 94:514–524
Torella D, Urbanek K, Rota M et al 2005 Local activation of cardiac stem cells repairs the infarcted heart and improves survival. Submitted
Urbanek K, Quaini F, Tasca G et al 2003 Intense myocyte formation from cardiac stem cells in human cardiac hypertrophy. Proc Natl Acad Sci USA 100:10440–10445
Wagers AJ, Weissman IL 2004 Plasticity of adult stem cells. Cell 116:639–648

DISCUSSION

Weissman: When we looked with anti-Sca-1 antibodies in human, we couldn't see any positive cells. It was only positive in certain mouse strains.

Nadal-Ginard: The antibody we use is produced by a Canadian company and recognizes Sca-1-positive cells from human, rat, dog and pig.

Weissman: When we started looking for human Sca-1 equivalents at that point we found out that it was a member of a big gene family, CD59. We had no way of knowing which one might be expressed on the cell surface. I am amazed at your results.

Nadal-Ginard: This antibody is very good. Unfortunately the company want to stop producing it. I am going to try to get them to give us the hybridoma line.

Weissman: Is it called anti-Sca-1 because they checked it is the same gene product, or because it sees similar cells?

Nadal-Ginard: They used Sca-1 as an immunogen and they checked this. We checked in human and compared the result with mouse.

Minger: It is my understanding that there are four different types of heart cells: atrial, mitral, ventricular and nodal cells. Do these progenitor cells or stem cells make all types?

Nadal-Ginard: Yes. These stem cells are pretty much interchangeable wherever they are isolated from. You can take the atrial cells and put them in the ventricle and you won't be able to tell the difference. It is difficult to do the reverse manipulation because the atrium is very thin.

Gepstein: What is the source of these cells? Are they are fetal remnant, or circulating stem cells that have remained there?

Nadal-Ginard: I don't know where they come from, but I have a suggestion. A few years ago we published a paper on human heart transplants. A female heart had been put into a man, and which was in there for only four days. After just four days we can find myocytes that are male in this female heart, and these are not fusions: they are male cells. This suggests to me that some of the cells can come from the circulation. I was intrigued by something that Irv Weissman published where he said that he did parabiosis and couldn't find anything. We did some parabiosis with a GFP male conjoined to a female. Two weeks later we find striated myocytes in the female that express GFP. There are quite a few of them, as many as 1 in 2000 myocytes. I suspect that some of those cells come from the circulation, and they are there long enough to have a phenotype that is different from that of a bone marrow cell.

Gepstein: What happens if you put these CSCs in a different tissue?

Nadal-Ginard: We are doing this now.

Goldman: You refer to the cells as being telomerase competent. Are they making telomerase, and what are the levels?

Nadal-Ginard: These cells have telomerase. Nevertheless, despite the fact that their levels are relatively high, the telomeres shorten with age. When these animals are given insulin-like growth factor (IGF), this telomerase shortening disappears and they become long again.

Soria: What was the composition of the growth factor cocktail you used?

Nadal-Ginard: It was exactly the same as the published growth factor cocktail for the neural stem cells.

Soria: Do these factors also work in a non-injured myocardium?

Nadal-Ginard: If you inject them they work in the sense that you get some more stem cells. In a normal healthy myocardium you get a slight increase in this type of cell. But we can't see any other changes.

Soria: Do you think there are local factors in the injured region that promote the regeneration?

Nadal-Ginard: Yes, I am sure that is correct. In fact, it is known that IGF goes up in the ischaemic zone. There are a number of other factors that seem to have a similar effect which are also increased. The amount that is made normally in the ischaemic zone doesn't seem to be enough to prevent the massive apoptosis that occurs. These cells have a feedback loop. When you give them IGF, they start making more IGF and they up-regulate the receptor. Once you get them going, they will go on for a month.

Soria: Has anyone tried to stop the scar formation?

Nadal-Ginard: If the scar isn't formed then you'll have a hole in the heart and a ruptured myocardium. This is not good: it causes sudden death.

Minger: I am intrigued by the culture conditions. Have you done this with human heart?

Nadal-Ginard: Yes.

Minger: Have you tried taking biopsies from humans and expanding cells?

Nadal-Ginard: We have tried to isolate them from two sources: surgical biopsies and donor or recipient hearts for transplants. For example, sometimes we have access to hearts that were going to be used for a transplant then for some reason aren't used, or a damaged heart that is taken out in a transplant operation.

Minger: How expandable are these cells?

Nadal-Ginard: From the rat and mouse they are very expandable (they have been grown for two years). I don't know about the human cells: this work is a lot more recent and so far they have gone to 40 or 50 doublings.

Weissman: In 1971 we published a paper on heart transplant patients from Stanford (Kennedy & Weissman 1971). This is pre-monoclonal antibodies. I made a deal with a heart transplanter that we could do an immediate autopsy on patients who had died of redevelopment of the coronary lesion. We found in one of three or four hearts one focus, where it was the HLA type of the host not the heart donor in a clear myeloproliferation.

Nadal-Ginard: People have been talking a lot about immunosuppressing animals to use other tissues. I showed a four day heart with a lot of chimerism. If this goes on at the same rate, you might suspect that in a year's time the heart should be the host heart, no longer in need of immunosuppression. However, this is wrong: all the chimerism happens in the first week after transplant and then stops dead. One of the reasons for this is that the immunosuppressive therapy stops this phenomenon, either because it stops the colonization or because it stops the differentiation of the cells. When you think about this therapy using immunosuppression, be aware that there may be a problem.

Björklund: Is there any connection between your cells and Guilio Cossu's mesangioblasts?

Nadal-Ginard: I think they are completely different in origin and spatial location. They might have some of the same effects.

Reference

Kennedy LJ Jr, Weissman IL 1971 Dual origin of intimal cells in cardiac-allograft arteriosclerosis. N Engl J Med 285:884–887

Generation of insulin-producing cells from stem cells

Bernat Soria*†, Enrique Roche*, Juan A. Reig* and Franz Martin*

**Institute of Bioengineering, University Miguel Hernandez, San Juan, 03550 Alicante, Spain, and †Department of Surgery, National University of Singapore, Singapore*

Abstract. Islet transplantation as a potential treatment for diabetes will always be limited mainly because of the difficulty in obtaining sufficiently large numbers of purified islets from cadaveric donors. One alternative to organ or tissue transplantation is the use of a renewable source of cells. Stem cells are clonogenic cells capable of both self-renewal and multilineage differentiation. Therefore, these cells have the potential to proliferate and differentiate into any type of cell and to be genetically modified *in vitro*, thus providing cells which can be isolated and used for transplantation. Moreover, these derived cells have proven to be useful in different animal models. In this regard, insulin-secreting cells derived from mouse embryonic stem cells normalize blood glucose when transplanted into streptozotocin-induced diabetic animals. Using a combination of several differentiation methods and a 'cell trapping' system, we have obtained insulin-secreting cells from undifferentiated embryonic stem cells. The construct used allows the expression of a neomycin selection system under the control of the regulatory regions of insulin gene and other β cell genes, such as *Nkx6.1*. Transplanted animals correct hyperglycaemia within 1 week and restore body weight in four weeks. Graft removal rescued the diabetic condition. Glucose tolerance test (IPGTT) and blood glucose normalization after a challenge meal was similar in control and in transplanted animals. This approach opens new possibilities for tissue transplantation in the treatment of type 1 and 2 diabetes.

2005 Stem cells: nuclear reprogramming and therapeutic applications. Wiley, Chichester (Novartis Foundation Symposium 265) p 158–173

Diabetes is a heterogeneous disorder affecting more that 150 million people in the world. Increasing prevalence, especially in Asian and Pacific countries, may reach 300 million people in the next 20 years (Zimmet et al 2001). Approximately 5% suffer type 1 diabetes, characterized by autoimmune destruction of the insulin-producing β cells of the pancreas, previously called juvenile diabetes because appears in young people and infants. Type 2 diabetes, is more related to lifestyle (sedentarism, obesity), usually appears in adults and represent 90% of diabetic patients. Non-insulin dependence is not the best description because there is a progressive loss in β-cell mass and 30–40% of these patients will need insulin injections. Disorders of either insulin action or insulin secretion may be the predominant feature. Both are usually present at the time this form of diabetes is clinically manifest. An increasing group of patients suffers MODY (maturity onset diabetes of the young), a group of monogenic diseases in which there is no autoimmune attack but a mutation in a key gene for β-cell physiology and/or development.

People who suffer from type 1 diabetes cannot produce insulin and will require external insulin administration throughout their life. Absolute lack of insulin is not compatible with life. Insulin is needed to control blood glucose and carbohydrate metabolism in the body. Although there are several hormonal mechanisms to increase blood glucose levels (glucagon, catecholamines, growth hormone, etc.), there is only one to decrease glucose (Soria 1997). Before the discovery of insulin, type 1 diabetes was an acute and life threatening disease. The discovery of insulin by Banting and Best in 1921 transformed type 1 diabetes into a chronic disease requiring daily controls of blood glucose and exogenous insulin injection. However, the poor control of blood glucose fluctuations afforded with this therapy often leads to severe complications including retinopathy, nephropathy, neuropathy, atherosclerosis and heart disease with a severe reduction of the quality and duration of life. A large multicentric study, the Diabetes Control and Complications Trial (1993), has recently shown that strict monitoring of blood glucose with intensive insulin therapy reduces the incidence of microvascular complications (nephropathy, retinopathy, etc). However, the resulting increased risk of hypoglycaemia and low patient compliance with these techniques preclude the widespread use of these techniques. Development of alternative therapies may reduce the incidence of complications and ameliorate the loss of quality of life in long-term diabetic patients.

Achieving normal glucose homeostasis would be a major advancement in the treatment of diabetes. Transplantation of islets from cadaveric donors as a potential treatment for diabetes has been investigated extensively over the past 30 years. Recent results from the Edmonton group have yielded, for the very first time, insulin-independence in type 1 diabetic patients (Shapiro et al 2000). However, such an approach will always be limited mainly because of the difficulty in obtaining sufficiently large numbers of purified islets. An estimation for Spain, which ranks number one in organ donation and has not a lower prevalence of diabetes, suggests that only 0.1% of the patients with diabetes could be treated in this manner. In order to make such a therapy easily available, new sources of insulin-producing cells must be identified. Furthermore, islet allotransplantation results in a permanent need for immunosuppression. Stem cell-derived therapies could be useful both in the supply of an unlimited source of cells and in inducing tolerance (Fändrich et al 2002).

Renewable cell sources for cell therapy of diabetes

There are three main renewable cell sources that can be used for β-cell replacement therapy: (i) xenoislets; (ii) tumour or transformed cell lines, and (iii) stem cells. Pig xeno-islets have been identified as the most suitable animal donor for islet transplantation, and reduction of the antigenicity of pig organs has been attempted by transgenic approaches. In addition to graft rejection, the risk from xenobiotic viruses is a further problem of xenotransplantation (Serup et al 2001). An alternative may be obtaining β-cell surrogates using tumoural or transformed cell lines which can be modified to express the defective genes. Nevertheless, the use of tumoural cell lines, either of pancreatic or neuroendocrine origin, in cell therapy of diabetes mellitus is restricted by several caveats (Soria et al 2000a, 2001). Tumoral cell lines retain their neoplastic nature displaying primitive differentiation features or even remnants of the parental cells. Furthermore, as a consequence of phenotypic instability and inadequate gene expression, most β-cell lines display maximum glucose responsiveness at much lower glucose concentrations than those observed in normal islets, and tend to lose insulin biosynthesis and regulated secretion with successive number of passages, making them useless in stable cell therapy treatments. It remains difficult to accept that cell engineering may result in insulin-producing cells with regulated insulin release. Pancreatic β-cells are unique in the way intracellular calcium is handled (Quesada et al 2000), ATP-dependent channels are regulated (Martin et al 1998) and neurohormonal regulation is performed (Nadal et al 2001). Both the glucose-sensing mechanisms and the secretory machinery have to be fully coordinated. In this sense, it may be very difficult to engineer β-cell surrogates which may be used in cell therapy (Bernà et al 2001, Soria et al 2000a).

Stem cells, which are clonogenic cells capable of both self-renewal and multilineage differentiation, may represent a clear alternative. Stem cells have the potential to proliferate, differentiate into many types of cells and to be genetically modified *in vitro*, thus providing cells which can be isolated and used for transplantation. In this regard, several groups have reported encouraging results showing the production of insulin-secreting cells from stem cells. Insulin-producing cells from stem cells have proven to be useful in normalizing blood glucose in streptozotocin diabetic animals (Soria et al 2000b).

Insulin-producing cells from embryonic stem cells

We have shown that insulin-producing cells may be obtained from mouse embryonic stem (ES) cells which were able to normalize glucose homeostasis when transplanted into an experimental model of diabetes mellitus (Soria et al

2000b). The strategy we used can be described in three steps (Soria et al 2001, Soria 2001): directed differentiation, cell-lineage selection and maturation. Afterwards, several studies appeared describing experiments in which mouse (Blyszczuk et al 2002, Hori et al 2002, Lumelsky et al 2001, Moritoh et al 2003, Vaca et al 2003) and human (Assady et al 2001) ES cells were differentiated into insulin-producing cells.

Mouse ES cells begin 'spontaneous' differentiation when they are deprived of leukaemia inhibitory factor (LIF), a cytokine that keeps mouse ES cells in the undifferentiated stage. Human ES cells do not respond to human recombinant LIF and it is much more tedious to keep them undifferentiated. The maintenance of pluripotency in human and mouse embryonic stem cells through activation of Wnt signalling by a pharmacological GSK3-specific inhibitor (Sato et al 2004) has been recently reported. Usually ES cells are cultured on a feeder layer (inactivated embryonic fibroblasts) that contribute with (unknown) messages to keep ES cells undifferentiated. In order to initiate differentiation and enter into the endodermal pathway it is better to form floating aggregates called 'embryoid bodies'. Under these conditions 'spontaneous differentiation begins and cells express markers of more differentiated cell types, for example the endodermal marker alfafetoprotein. The differentiation process begins with non-differentiated pluripotent mouse ES cells that are cultured in the absence of LIF and forced to form floating aggregates (EB) when cultured in bacterial plates or in hanging drops. EB reproduce some of the conditions necessary for cell differentiation: (i) polarization; (ii) cell-to-cell interactions; (iii) presence of asymmetric contacts and (iv) production of growth factors. For example, ENS ES cells were cultured for 21 days forming EB, and expressed transcripts of endocrine (i.e. insulin, glucagon and PP) and exocrine- (i.e. amylase, elastase and carboxypeptidase) specific genes (Skoudy et al 2004, Soria et al 2001), suggesting that transcriptional factors involved in their regulation are functional in these ES cells. Assady et al (2001) reported that removal of mouse embryonic fibroblasts and transfer to suspension culture in human ES cells (H9) is enough to observe expression of the insulin gene and other islet specific genes (glucokinase, Glut-2) with simultaneous decrease in *Oct-4* expression. However, these results have not been replicated by other authors and may represent a H9 cell line-specific feature. In other assays insulin-producing cells do not appear spontaneously and factors have to be added to the culture media ('directed differentiation'). In our initial report we used nicotinamide (10 mM) to increase generation of insulin-producing cells. By a yet unknown mechanism, nicotinamide induces differentiation of fetal islet (Otonkoski et al 1993) ES cells (Soria et al 2001b, Vaca et al 2003) into insulin-producing cells. This observation has been confirmed by other authors. The approach of Lumelsky et al (2001) consisted of multistep culture conditions, based on those used for generating neural cells, which began with the production of an enriched population of nestin-positive cells. After expansion with basic fibroblast growth factor (bFGF or FGF2), B27 supplement and nicotinamide were added to promote differentiation. Hori et al (2002) also used a multistep culture protocol, very similar to the previous one, but with the addition of the inhibitor of PI3K, LY294002 in the last step. Blyszczuk et al (2002) forced the differentiation by the constitutive expression of the *Pax4* gene and added a cocktail of factors in the last step (insulin, transferrin, progesterone, putrescine, Na-selenite, laminin, fibronectin, B27 supplement and nicotinamide). Finally, Moritoh et al (2003) also devised a multistep approach in which a cocktail of factors were added during the last step. The differentiation soup contains the same factors that were used by Blyszczuk and colleagues, with the exception of laminin and fibronectin. In contrast keratinocyte growth factor, epidermal growth factor and FGF2 were added. Most of these factors are in a long list published by one of us in 2001 (Soria 2001). Based on the idea that *in vitro* differentiation may recapitulate some of the steps of *in vivo* development, the majority of the maturation protocols are based on the knowledge gained on studies from fetal islets, rat pancreatic regeneration and adult pancreatic duct cells. We have assayed different maturation strategies (sodium butyrate, mytomicin C, nicotinamide, low and high glucose, LY294002) with varied results on cell proliferation and differentiation. For example, sodium butyrate was effective increasing insulin content, but most of the cells died in the process. Taking these results together, it is reasonable to postulate that a sequential and/or combined exposure to growth factors, extracellular matrix components, nutrients and culture conditions may end with a high proportion of cells that express and retain the β-cell phenotype. A better knowledge of β-cell development, together with prospective identification and purification of suitable precursor cells, will help to drive the obtention of better insulin-producing cells. Embryonic germ cells have the same pluripotentiality as ES cells. In spite of this, up to now, no reports have appeared showing the possibility of using embryonic germ cells to obtain insulin-producing cells.

However, none of these methods generates a pure population of the desired cell type and it is mandatory to find methods that allow us to select one cell type and discard the other (usually unknown) cell types. The cell-trapping strategy that we introduced is based on the expression of a candidate gene, for example insulin. In summary, cells are transfected with a gene that fuses the regulatory region of the insulin gene with βgeo, a chimeric gene which encodes for both β-galactosidase activity and neomycin resistance. This allows the selection of insulin-expressing cells when in culture with neomycin, because they also express the neomycin resistance gene. Furthermore, expression of β-galactosidase makes possible the histochemical identification of transplanted cells. This cell lineage selection system allows multiple variants, such as using the promoter region of other markers, for example Pdx-1, neurogenin 3, PP and HNF1β (Bernà et al 2001). A safety procedure to protect against unpredicted malignancies can be added. For example, ligating the Herpes tymidine kinase gene to the construct makes it possible to suppress cells by treating with gancyclovir.

Rajagopal et al (2003) have recently reported that ES cells may uptake insulin that is usually present in the culture media. If this is the case, then insulin staining is an insufficient criterion. Furthermore, adsorbed insulin is retained for more than three weeks in insulin-deficient media. These authors showed that the Lumelsky et al (2001) protocol generates apoptotic cells with no insulin 1 gene expression and very low insulin 2 gene expression, a very low insulin content (0.02% of mouse islet cells), no granules detected in electron microscopy, etc. These observations necessitate the use of more restrictive criteria, including C-peptide, presence of granules, *in situ* hybridization, Northern analysis and transplantation for longer periods. Table 1 summarizes which criteria have been fulfilled so far.

Insulin-producing cells from adult stem cells

It has been generally accepted that terminally differentiated cells have suspended their differentiation possibilities and maintain a fixed phenotype with negligible division potential, whilst other tissues, such as skin, intestinal epithelia or blood cells, retain the ability to regenerate throughout life due to the presence of stem cells. A third category are those tissues such as liver and muscle, that may regenerate when injured or as a consequence of exercise. It is known that β cell mass can increase under certain circumstances (pregnancy, incipient obesity) and experimental pancreatic endocrine regeneration has been demonstrated during or after transient hyperglycaemia, ductal ligation, partial pancreatectomy, cellophane wrapping, etc. Furthermore, it has been estimated that the lifespan of a β cell is between 1–3 months. However, the identification of an adult progenitor for islet cells is still a matter of debate.

The proposed candidates for islet regeneration are:

- nestin-positive cells isolated from rat and human islets (Zulewski et al 2001)
- intra-islet precursor cells expressing somatostatin and the transcription factor PDX-1 (pancreatic duodenum homeobox) (Ramiya et al 2000) and
- ductal cells (Bonner-Weir et al 2000).

Bonner-Weir at al (2000) have demonstrated that duct tissue from human pancreas forms islet-like structures when cultured on Matrigel in the presence of ITS medium (insulin, transferrin, selenium). Ductal cells also initiate β cell phenotype when they are virally transfected with the early islet developmental transcription factor neurogenin 3 or PDX1 (Noguchi et al 2003). Nevertheless, the low expansion capability of these procedures has so far precluded their clinical application.

TABLE 1 Criteria fulfilled so far by insulin-producing cells derived from embryonic (ES) and adult stem cells

	ES cells	*Adult stem cells*
Insulin content	3, 5, 8, 10	
Glucagon	3, 5	
Insulin release	3, 8, 10	11
C-peptide content	3	11
Secretory granules		4, 11
Blood glucose	2, 3, 8	4, 11
Normalization		
– Graft withdrawal	3	
– Glucose tolerance	3, 8	
X-gal	6, 9	
Immunocytochemistry		
– Insulin	1, 2, 5, 6, 10	
– Glucagon	2, 3, 5, 6	
– Glut-2	3	
– Glucokinase	3	
– Pdx-1	3	4
– C-peptide	3	
Gene expression		
– Insulin 1	3, 5, 6	4
– Insulin 2	5, 6	4
– Hum insulin	1	
– Glucokinase	1, 6	
– SUR–1	1, 5	
– Neurogenin3	1, 3	4
– Pdx1	2, 7	11
– Nkx6.1		4,11
– Kir6.2		6
– Neuro D		4
– Pax 6		4
– Pax 4	2	4
– Isl 1		4

References: (1) Assady et al 2001, (2) Blyszczuk et al 2002, (3) Hori et al 2002, (4) Kojima et al 2003, (5) Lumelsky et al 2001, (6) Moritoh et al 2003, (7) Skoudy et al 2004, (8) Soria et al 2000b, (9) Soria et al 2000a (10) Vaca et al 2003, (11) Zalzman et al 2003.

Stem cells from extrapancreatic sources have also been postulated to generate insulin-producing cells. Adult hepatic oval stem cells are reputed to differentiate into insulin-producing cells both *in vitro* (Yang et al 2002) and *in vivo* (Kojima et al 2003, Zalman et al 2003). Moreover, there is increasing evidence that both bone marrow stem cells and peripheral-blood stem cells (Krause et al 2001) have multiple differentiation potential into numerous tissue types and organs. Two recent reports

(Hess et al 2003, Ianus et al 2003) have shown that bone marrow may contribute to islet regeneration. In contrast, another report (Choi et al 2003) suggests that bone marrow-derived cells are a distinct population from islet cells and that transdifferentiation from bone marrow-derived cells to pancreatic β cells is rarely observed. However, any direct evidence that bone marrow-derived cells contributed to repopulating human islets is lacking.

Conclusion

Regeneration therapy for pancreatic β cells includes direct replacement with cells obtained from cadaveric donors (islet transplantation) or differentiated from stem cells, in conjunction with immunosuppressive therapy or immuno-isolation. In a second strategy, *ex vivo* regeneration therapy, a patient's own cells, such as bone marrow stem cells, are transiently removed and induced to differentiate into β cells *in vitro*. A third category for islet regeneration from stem cells will include the use of factors that may stimulate the *in vivo* capability of islet regeneration, especially in type 2 diabetic patients. It is hoped that these strategies will, in the future, provide a cure for diabetes, but only the first (islet transplantation) is now available in the clinics.

At present it is difficult to decide which approach is the most promising. Autoimmune profiles could benefit from tolerance induction by stem cells, whilst monogenic forms of diabetes (MODY) may need *ex vivo* gene therapy of patient-derived cells and type 2 diabetic people may stop the progress of β-cell destruction stimulating pancreas regeneration. Islet cell and stem cell research is set to develop rapidly. Potential advantages of ES cells are that: (i) they can be cultured for generations and (ii) in theory, they could be engineered to express the appropriate genes that would allow them to escape or reduce detection by the immune system. Nevertheless, most researchers agree that all potential sources of islet cells should be extensively investigated, including both adult and embryonic stem cells.

Acknowledgements

These studies have been supported by the European Union (QLK3-CT-2002-0177), Ministry of Science and Technology (PM99-0142; SAF 2003-03307; GEN2001-4748-C05-05), Instituto de Salud Carlos III (C03-171; C03-212; C03-210) and the European Foundation for the Study of Diabetes.

References

Assady S, Maor G, Amit M, Itskovitz-Eldor J, Skorecki KL, Tzukerman M 2001 Insulin production by human embryonic stem cells. Diabetes 50:1691–1697

Bernà G, León-Quinto T, Enseñat-Waser R, Montanya E, Martin F, Soria B 2001 Stem cells and diabetes. Biomed Pharmacother 55:206–212.

Blyszczuk P, Czyz J, Kania G et al 2002 Expression of Pax4 in embryonic stem cells promotes differentiation of nestin-positive progenitor and insulin-producing cells. Proc Natl Acad Sci USA 100:998–1003

Bonner-Weir S, Taneja M, Weir GC et al 2000 'In vitro' cultivation of human islets from expanded ductal tissue. Proc Natl Acad Sci USA 97:7999–8004

Choi JB, Uchino H, Azuma K et al 2003 Little evidence of transdifferentiation of bone marrow-derived cells into pancreatic beta cells. Diabetología 46:1366–1374

Diabetes Control and Complications Trial 1993 The effect of intensive treatment of diabetes on the devlopment and progression of long-term complications in insulin-dependent diabetes mellitus. N Engl J Med 329:977–986

Fändrich F, Lin X, Chai GX et al 2002 Preimplantation-stage stem cells induce long-term allogenic graft acceptance without supplementary host conditioning. Nat Med 8:171–178

Hori Y, Rulifson IC, Tsai B, Heit JJ, Cahoy JD, Kin SK 2002 Growth inhibitors promote differentiation of insulin-producing tissue from embryonic stem cells. Proc Natl Acad Sci USA 99:16105–16110

Hess D, Li L, Martin M et al 2003 Bone marrow-derived stem cells initiate pancreatic regeneration. Nat Biotechnol 21:763–770

Ianus A, Holz GG, Theise ND, Hussain MA 2003 In vivo derivation of glucose-competent pancreatic endocrine cells from bone marrow without evidence of cell fusion. J Clin Invest 111:843–850

Kojima H, Fujimiya M, Matsumura K et al 2003 NeuroD-betacellulin gene therapy induces islet neogenesis in the liver and reverses diabetes in mice. Nat Med 9:596–603

Krause DS, Theise ND, Collector MI et al 2001 Multi-organ, multi-lineage engraftment by a single bone marrow-derived stem cell. Cell 105:369–377

Lumelsky N, Blondel O, Laeng P, Velasco I, Ravin R, McKay R 2001 Differentiation of embryonic stem cells to insulin-secreting structures similar to pancreatic islets. Science 292:1389–1394

Martin F, Pintor J, Rovira JM, Ripoll C, Miras-Portugal MT, Soria B 1998 Intracellular diadenosine polyphosphate: a novel second messenger in stimulus-secretion coupling. FASEB J 12:1499–1506

Moritoh Y, Yamato E, Yasui Y, Miyazaki S, Miyazaki J 2003 Analysis of insulin-producing cells during in vitro differentiation from feeder-free embryonic stem cells. Diabetes 52:1163–1168

Nadal A, Ropero AB, Laribi O, Maillet M, Fuentes E, Soria B 2000 Non-genomic action of estrogens and xenoestrogens by binding at a plasma membrane receptor unrelated to estrogen receptor alfa and estrogen receptor beta. Proc Natl Acad Sci USA 97:11603–11608

Noguchi H, Kaneto H, Weir GC, Bonner-Weir S 2003 PDX-1 protein containing its own antennapedia-like protein transduction domain can transducer pancreatic duct and islet cells. Diabetes 52:1732–1737

Otonkoski T, Beattie GM, Mally MI, Ricordi C, Hayek A 1993 Nicotinamide is a potent inducer of endocrine differentiation in cultured human fetal pancreatic cells. J Clin Invest 92:1459–1466

Quesada I, Martín F, Soria B 2000 Nutrient modulation of polarized and sustained submembrane Ca^{2+} microgradients in mouse pancreatic islet cells. J Physiol (Lond) 525:159–167

Rajagopal J, Andersson WJ, Kume S, Martinez OL, Melton DA 2003 Insulin staining of ES cell progeny from insulin uptake. Science 299:363

Ramiya VK, Maraist M, Arfors KE et al 2000 Reversal of insulin-dependent diabetes using islets generated 'in vitro' from pancreatic stem cells. Nat Med 6:278–282

Sato N, Meijer L, Skaltsounis L, Greengard P, Brivanlou AH 2004 Maintenance of pluripotency in human and mouse embryonic stem cells through activation of Wnt signalling by a pharmacological GSK-3-specific inhibitor. Nat Med 10:55–63

Serup P, Madsen OD, Mandrup-Poulsen T 2001 Islet and stem cell transplantation for treating diabetes. Br Med J 322:29–32

Shapiro AMJ, Lakey JRT, Ryan EA et al 2000 Islet transplantation in seven patients with Type 1 Diabetes Mellitus using a glucocorticoid-free immunosuppressive regimen. N Engl J Med 343:230–238

Skoudy A, Rovira M, Savatier P et al 2004 TGFbeta, FGF and retinoid signalling pathways promote pancreatic exocrine gene expression in mouse embryonic stem cells. Biochem J 379:749–756

Soria B 1997 Physiology and pathophysiology of the islets of Langerhans. Advances in Experimental Medical Biology, Vol 426. Plenum Press, New York

Soria B 2001 In vitro differentiation of pancreatic β-cells. Differentiation 68:205–219

Soria B, Andreu E, Bernà G et al 2000a Engineering pancreatic islets. Pflügers Arch 440:1–18

Soria B, Roche E, Bernà G, León-Quinto T, Reig JA, Martin F 2000b Insulin-secreting cells derived from embryonic stem cells normalize glycemia in streptozotocin-induced diabetic mice. Diabetes 49:157–162

Soria B, Skoudy A, Martin F 2001 From stem cells to beta cells: new strategies in cell therapy of diabetes mellitus. Diabetologia 44:407–415

Vaca P, Berna G, Martin F, Soria B 2003 Nicotinamide induces both proliferation and differentiation of embryonic stem cells into insulin-producing cells. Transplant Proc 35:2021–2023

Yang L, Li S, Hatch H et al 2002 'In vitro' trans-differentiation of adult hepatic stem cells into pancreatic endocrine hormone-producing cells. Proc Natl Acad Sci USA 99:8078–8083

Zalzman M, Gupta S, Girl RK et al 2003 Reversal of hyperglycemia in mice by using human expandable insulin-producing cells differentiated from fetal liver progenitor cells. Proc Natl Acad Sci USA 100:7253–7258

Zimmet P, Alberti KG, Shaw J 2001 Global and societal implications of diabetes epidemic. Nature 414:782–787

Zulewski H, Abraham EJ, Gerlach MJ et al 2001 Multipotential nestin-positive cells isolated from adult pancreatic islets differentiate *ex vivo* into pancreatic endocrine, exocrine and hepatic phenotypes. Diabetes 50:521–533

DISCUSSION

Gearhart: What is known about the autoimmunity—the basis of the disease? Are you ignoring the autoimmunity and just putting more cells in?

Soria: There are two types of diabetes, type 1 and type 2. It was accepted that type 1 was autoimmune, but now there is more evidence that a high proportion of cases aren't. Let's say that 30–40% of cases are MODY (maturity onset diabetes of the young). This is caused by a single gene mutation and not by autoimmunity. In islet transplantation we have to use immunosuppressors because rejection must be avoided as well as autoimmunity. Even if you get the cells from the same individual by means of therapeutic cloning, you still need to solve the autoimmunity if this is an autoimmune disease. The process of autoimmunity is related to GAD. No one has a clear explanation of the relationship between

GABA and autoimmunity, but these cells release GABA and also the GABAergic cells in the brain have insulin. There is some association. This is a protein that appears in the membrane and is attacked, but it is not the only target for autoimmune disease. This is a complex situation. Another approach that has been suggested has been to stop autoimmunity, for example by giving immunosuppressants during the first stages of diabetes. There are no clear results yet. I think islet transplantation is very useful for diabetes, but in at least half of the type 1 diabetic patients we also need to solve autoimmune disease. In the other half it is caused by a mutation, which we will need to address. Alternatively we could give an islet from a different genetic source and mix it with endothelial progenitor cells from the host, which is a way of reducing the immune attack. Type 2 diabetes may be the best option because this is not autoimmune.

Grompe: If you had a large source of islets, you could do repeat transplants. It is not that big a procedure to go into the hepatic artery and transplant more cells. I don't understand why this has to be a one time procedure. It could be possible that islets don't self-renew.

Soria: That would be fine, provided you have enough cells. In fact, islet transplantation works better if you proceed with a subsequent islet transplant.

Gepstein: You told us that the pancreatic islet has to be a functioning unit in order to secrete insulin in a proper manner. Are the cells in the aggregates already in contact? Do you see connexins in the cells?

Soria: I don't have a clear answer yet. I couldn't show that the aggregates work like an islet. This is something we need to explore more carefully.

Gepstein: People were talking about doing encapsulations to prevent autoimmunity. What has happened with this?

Soria: Encapsulation is a nice idea, but the problem with it is that the cut-off window is around 1 kDa. This is big enough for insulin, but it also the size of the cytokines that kill the islet and enter through the same pathway. The results so far with transplanting encapsulated pig islets into humans have been unsuccessful.

Wilmut: What is the cause of type 2 diabetes, and would it be treated by transfer of islet cells?

Soria: Type 2 diabetes is associated with lifestyle, obesity and sedentarism. With incipient obesity people increase their β-cell mass because more insulin is needed, but increase in the β-cell mass is very limited, additionally when a certain level of obesity is reached and there is no exercise, insulin resistance appears. Then the cells become exhausted and die, and also high glucose and high lipid levels kill the cells. When type 2 diabetes is diagnosed, within a few years the β-cell mass of the patient gets very low. At this stage, transplantation of β-cells may help together with other therapeutic strategies.

Wilmut: Could you say more about insulin resistance?

Soria: Type 2 diabetes patients have both defects in insulin release and also insulin resistance. This means that they need more insulin to have the same effect. We have some drugs that may increase the response to insulin. A combination of these drugs together with an increase in the β-cell mass may help recover the normal situation. Insulin resistance is a consequence of obesity, dislipaemia and sedentarism.

Wilmut: Is cell transplantation used effectively in type 2 diabetes? It sounds as if you need to get the patient to change their lifestyle.

Soria: In the most common form of type 2 diabetes islet transplantation may help to recover β-cell mass, but lifestyle also needs to be changed. This is true for caucasians, but I am curious about type 2 diabetic patients in Asia, who are not necessarily obese.

Solter: Is cell transplantation done at all today?

Soria: It has been done. There is a group in Austria which has transplanted at least one series of type 2 diabetic patients. Type 2 is as complex and devastating as type 1 in the end. It is associated with blindness, kidney failure and so on. The absolute need for insulin in type 1 diabetic patients has overshadowed the ill effects of type 2. In addition, type 2 represents over 90% of diabetic people. If we had enough islet cells for everyone, type 2 patients could have transplants as well.

Verfaillie: Are the islets you are creating from ES cells which express insulin and glucagon the very early double-positive cells that occur during development?

Soria: Yes. Some people have reported that the progenitors can stain for both insulin and glucagons.

Verfaillie: Can you drive them to become single positive *in vitro*?

Soria: Not yet. In terms of staining, 100% of cells are positive for insulin and 88% are positive for glucagon. But whilst insulin content is between 10 and 20% of controls, glucagon content is only 1% of the controls (P. Vaca et al, unpublished results). This means that most of the protein they have is insulin and not glucagon. This is one thing we want to check when we remove the grafts, to see whether ES derived cells mature inside the animal.

Trounson: One of the strategies you spoke of was to overexpress genes. I am not sure that the regulatory agencies will accept this for human therapies. In the case of overexpressing genes you have to be sure that at some stage you silence them. Is this the kind of strategy that you have in mind? And why is β-cell derivation via the liver cell better than selecting directly for a pancreatic phenotype?

Soria: Gene expression, in developmental biology at least, is a very powerful strategy. *In vitro* it doesn't work as well. If you express the transcription factor Pdx1 you don't get insulin-producing cells. However, constitutive expression of Pax4 in ES cells generates insulin-producing cells (Blyszczuk et al 2003). With liver cells it seems is different. Why do we target liver? Liver is also a derivative of the endoderm and in certain aspects is very close to the pancreas tissue. In *in vivo* experiments transfection of liver (stem?) cells with NeuroD and β cellulin has been used to generate insulin-producing cells (Kojima et al 2003). This strategy may be combined with the embryonic stem cell methods to induce differentiation into the endoderm.

Nadal-Ginard: What is the normal turnover of the islets?

Soria: In the mouse it is less than 1% over 24 h. To our knowledge the mouse pancreas regenerates, but the human pancreas doesn't. No one has shown human islet regeneration. The mouse will recover from a partial pancreatectomy in two weeks.

Nadal-Ginard: If the mouse pancreas turns over cells this rapidly, it means it must have very active regeneration. Isn't it worth looking for the progenitors in the pancreas to see whether you can discover them?

Soria: It was already known that the β-cell mass may increase during pregnancy and incipient obesity and most of the previous work indicates that epithelial cells from the pancreatic ducts are the progenitors for islet neogenesis. However, a recent report by Doug Melton's group has challenged this view demonstrating that replication of mature β cells are responsible for the increase in β-cell mass (Dor et al 2004).

Goldman: You mentioned selection strategies involving Nkx6.1 and neurogenin 3. Using these normally serially acquired markers, can you get different phenotypes or different ratios of α and β cells?

Soria: I can't answer that. I got most of the data with two regions of the promoter of the insulin gene, and also Nkx6.1. In the case of Nkx6.1 the phenotype is very close to that of the cells that I presented here. In the case of Pdx1 and neurogenin 3, I am experiencing problems. This is another difficulty with moving to human ES cells: every time a referee asks for another experiment it means six months' more work. With human ES cells it will take longer because they don't grow so fast as mouse ES cells.

Grompe: I really like your emphasis on β cell physiology, and that insulin alone is insufficient. You talked about certain functions that you would like these cells to have, such as the K^+ channel regulated release. We always hear about positive factors when we discuss ES cell differentiation. What kind of genes should not be switched on in these kinds of cells? Let's say that a cell releases insulin, responds to amino acids and so on: are there some genes that you would consider to be potentially dangerous? Would it be bad, for example, if these stem cells made secretin?

Soria: Do you mean the differentiated cells or the cells in the process of differentiation?

Grompe: I am talking about the final cell: what do you screen for to make sure they don't make it?

Soria: That is a good question, but there is no clear answer. Let's say that we need a cell that behaves like a β cell and that has three groups of proteins: the glucose-sensing machinery (Glut-2, glucokinase and the ATP-regulated K^+ channel), the insulin synthesis and processing machinery (pre-proinsulin, convertases) and the exocytotic machinery (secretory granules, syntaxin, synaptotagmin, etc). Insulin is expressed in other tissues, for example in the brain. In the mouse we can check for Insulin 1 (present only in the islets) and Insulin 2 (both in islets and brain). We may also check whether neural genes are expressed (GFAP, neurofilament 200, Tuj1, acetylcholinesterase, tyrosinhydroxylase, myelin basic protein, etc.). In addition, cell physiology is the appropriate way to explore whether these proteins are functional.

Grompe: Or markers for the exocrine pancreas. You don't want to transplant a cell that then makes trypsin or chymotrypsin.

Soria: That is something that we know. We have a strategy for going into the exocrine pathway, and we are pretty sure that we are not in the exocrine pathway with these cells (Skoudy et al 2004). However, even if we transplant some exocrine tissue, it doesn't really matter as long as this contains differentiated cells.

Minger: We haven't really discussed this, but it is clear from data by Peter Jones from King's College London that having cells that make insulin alone is insufficient. You have to configure these cells in some way, such as into islets or pseudo-islets. Individual cells that make insulin don't respond to glucose appropriately. We can make the cells, but then there is a tissue engineering component to this that is very difficult. Type 1 diabetes is a somewhat different situation to Parkinson's disease where if we put in enough single dopamine producing cells this might be sufficient.

Soria: I agree. Engineering an islet is the next step. One solution would be to go into the progenitor (Pdx1- or neurogenin 3-positive cells) and drive those cells through the last differentiation steps. Another would be to make matrices that would help to build this 3D structure.

Nadal-Ginard: How many cells do you need to correct diabetes?

Soria: From clinical islet transplantation we know that a patient that weighs 70 kg needs 800 000 islets, and each islet has a minimum of 1000 cells, this results in 800 million cells, but most probably a percentage of these cells have been damaged during the isolation process. If 50% of the cells are not functional that will end with 400 million cells. On the other side, we have one million islets in our pancreas, each islet may have between 1000 and 2000 cells, but type 1 diabetes (loss of function) is diagnosed when 90% of the β-cell mass has been lost. This gives a figure of 100 to 200 million cells. The response for a patient of 70 kg will be between 200 and 400 million cells.

Trounson: Are you certain you need an islet? Or could you get glucose responsive insulin producing cells to work elsewhere?

Soria: An aggregate of β cells is better that intensive insulin therapy and an islet, that combines β and non-β cells will be better that an aggregate of β cells. Human islets have been transplanted into the liver. We know from clinical islet transplantation that islets are happy in the liver. This works. If we are talking about animal experimentation, the only data are from attempts to convert liver stem cells into insulin-producing cells. In this case they grow in small colonies.

Grompe: In a paper by Larry Chan's group they claim to be able to correct chemically-induced diabetes by just doing sequential gene transfer with the adenoviral vectors into the liver (Kojima et al 2003). But they really didn't describe what cells became insulin positive and whether they were formed into islets or not.

Soria: Electron microscopy of these cells showed that they look similar to islet cells.

Minger: We can put Pdx1 into both rat and human liver and get Pdx1 expression, yet not detect insulin at all. We have replicated this several times.

Soria: This was published by Sara Ferber in 2000 (Ferber et al 2000) but it has proved to be difficult to replicate. In a more recent paper Zalzman et al (2003) were able to generate insulin-producing cells from liver stem cells.

Grompe: We have reproduced it. It is not Pdx1 alone, but Pdx followed by neurogenin 3. Pdx1 induces massive hepatic damage. This is why Larry Chan used an anti-apoptotic gene product as well.

Nadal-Ginard: If you take type 1 diabetes and treat it with insulin, for the first few years you have cured the disease. But 10–15 years later you start seeing complications. How are you going to get round this with cell therapy?

Soria: We have experience with islet transplantation. The experience is limited but over the last 25 years over 500 people have been transplanted. Only over the last 4 years have the protocols been better, but from those people that were transplanted 25 years ago we learned that nephropathy is reversible. In contrast, the retinal damage is not recoverable. After islet transplantation, the progressive damage of certain tissues stops and the quality of life is much improved. The first target is therefore diabetic people with lots of complications.

Gepstein: Were the pancreatic buds experiments done using conditioned media from embryonic tissues?

Soria: We have done both. In the first series of experiments differentiation was induced by co-culture of mouse ES cells with pancreatic buds from E14–E17 embryos (Leon-Quinto et al 2004). In more recent experiments conditioned media from E14–E17 pancreatic buds was used with similar results.

Gepstein: Does the ES differentiation increase only the beta cells?

Soria: I don't know, because I select for cells that produce insulin.

Gepstein: What happens without selection?

Soria: If you don't have conditioned media at the end of the protocol you could have 2–3% of the cells positive for insulin. With conditioned media we have 20–30% positive. I don't know whether these cells also produce other pancreatic cell types because of the selection criteria we use.

Minger: Are the cells expandable?

Soria: That is an important point. It depends on the protocol. The differentiation could be 20–30% of the control. If you culture the cells for two more weeks *in vitro*, then they don't maintain the phenotype. I cannot say that the phenotype is the same. Even with cells that do proliferate *in vitro*, when you transplant the cells into the kidney or spleen, they do not produce tumours. We do not see tumours with selected cells but we do with cells that have not been selected previously.

References

Blyszczuk P, Czyz J, Kania G et al 2003 Expression of Pax4 in embryonic stem cells promote differentiation of nestin-positive progenitor and insulin-producing cells. Proc Natl Acad Sci USA 100:998–1003

Dor Y, Brown J, Martinez OI, Melton DA 2004 Adult pancreatic beta-cells are formed by self-duplication rather than stem-cell differentiation. Nature 429:41–46

Ferber S, Halkin A, Cohen H 2000 Pancreatic and duodenal homeobox gene 1 induces expression of insulin genes in liver and ameliorates streptozotocin-induced hyperglycemia. Nat Med 6:568–572

Kojima H, Fujimiya M, Matsumura K et al 2003 NeuroD-β cellulin gene therapy induces islet neogenesis in the liver and reverses diabetes in mice. Nat Med 9:596–603

Leon-Quinto T, Jones J, Skoudy A, Burcin M, Soria B 2004 In vitro directed differentiation of mouse embryonic stem cells into insulin-producing cells. Diabetologia 47:1442–1451

Skoudy A, Rovira M, Savatier P et al 2004 TGF-β, FGF and retinoid signalling pathways promote pancreatic exocrine gene expression in mouse embryonic stem cells. Biochem J 379:749–756

Zalzman M, Gupta S, Girl RK et al 2003 Reversal of hyperglycemia in mice by using human expandable insulin-producing cells differentiated from fetal liver progenitor cells. Proc Natl Acad Sci USA 100:7253–7258

Cell therapy for Parkinson's disease: problems and prospects

Anders Björklund

Wallenberg Neuroscience Center, Division of Neurobiology, Lund University, BMC A11, S 22184 Lund, Sweden

Abstract. Cell replacement therapy in Parkinson's disease (PD) has so far been based on the use of primary dopaminergic (DA) neuroblasts obtained from the brain of aborted human fetuses. Clinical trials show that intrastriatal DA neuron transplants can give substantial symptomatic relief in advanced PD patients. Two recent NIH-sponsored placebo-controlled trials, however, have given disappointing results and highlighted a number of critical issues that need to be resolved in order to turn cell transplantation into an acceptable clinical therapy. *First*, graft survival and clinical outcome has so far been too variable, suggesting that DA neuron grafts may not be equally effective in all PD patients. *Secondly*, it has become clear that immune mechanisms leading to slowly developing inflammatory responses may compromise long-term graft survival and function. *Third*, the problems associated with the use of tissue from aborted fetuses make it necessary to develop alternative sources of cells for transplantation. Recent progress in the generation of DA neuroblasts from neural progenitors and embryonic stem cells suggest that these kinds of cell may offer more accessible, defined and standardized sources of cells for clinical transplantation in PD.

2005 Stem cells: nuclear reprogramming and therapeutic applications. Wiley, Chichester (Novartis Foundation Symposium 265) p 174–187

Parkinsons's disease (PD) is one of the most interesting potential neurological disease targets for restorative cell therapy. Animal experimental studies carried out over the last 25 years have provided substantial evidence that this approach may be feasible. Studies in animal models of PD, where the nigrostriatal dopamine (DA) system is destroyed either by the 6-hydroxydopamine (6-OHDA) toxin (in rats) or by MPTP (1-methyl-4-phenyl-1,2,3,6-tetrahydropyridine) (in monkeys) have shown that fetal dopaminergic neuroblasts obtained from the developing ventral mesencephalon (VM) of mid-gestational rat or mouse fetuses, or from the developing nervous system of 6–8 week old aborted human fetuses, survive transplantation to the striatum and can reverse some, but not all, of the motor symptoms associated with the loss of nigral DA neurons (see Björklund 1992, Winkler et al 2000, for comprehensive reviews).

The functional effect develops slowly: the first transplant-induced effects on motor behaviour are seen after 3–6 weeks with grafts of fetal rodent VM cells, and after a few months with fetal human DA neurons, and are correlated with the growth of axonal processes from the grafted cells and the establishment of a new DA-containing terminal network in the host striatum. Rodent studies have shown that the grafted DA neurons are spontaneously active and will restore a tonic, baseline release of DA to near-normal levels within the areas reinnervated by the grafted cells. In 6-OHDA-lesioned rats, functionally active intrastriatal VM grafts, which contain between 1000 and 2000 surviving DA neurons (about 10–15% of the number of DA neurons normally present in the substantia nigra) have been shown to restore striatal DA to about 10–20% of normal levels, *in vivo* striatal synthesis by about 20–30%, and striatal DA release (as measured by microdialysis in the area densely innervated by the grafted neurons) by 40–100% of normal levels (see Björklund 1992, Winkler et al 2000, Brundin & Hagell 2002, for references).

These data indicate that grafted DA neurons are tonically active and that they synthesize and release their transmitter at a rate that may be even higher than that of the normal nigrostriatal afferents. Using multiple graft placements it has been possible to obtain a striatal DA innervation density in the range of 40–75% that of normal (as assessed by tyrosine hydroxylase (TH) immunoreactivity; Nikkhah et al 1994, Winkler et al 1999, 2000, Kirik et al 2001). Also in these cases, however, functional recovery has been incomplete: motor asymmetry, as assessed by drug-induced turning behaviour, is completely reversed, while several aspects of spontaneous motor behaviour, such as stepping, forelimb use and postural balance, are only partly restored.

There are several reasons why functional recovery is incomplete. One obvious limitation of the current grafting procedure is that the DA neuroblasts are placed ectopically, i.e. in the striatum rather than in the substantia nigra where they normally are located. Some of the major afferent inputs that normally play a role in the phasic regulation of the nigrostriatal neurons are thus not available to the grafted cells. This lack of afferent control may limit the ability of the grafted DA neurons to restore more complex motor behaviours. An alternative explanation is that the graft-derived innervation is confined to the striatum while the denervating lesion also involves other non-striatal areas. The inability of the intrastriatal VM grafts to provide complete functional recovery may thus be due to the fact that limbic and cortical areas, as well as other parts of the basal ganglia, remain denervated and thus unaffected by the transplanted cells. In support of this view we have shown that graft-reduced functional recovery is more pronounced in animals with partial lesions of the nigrostriatal pathway, in which the denervation is confined to the striatum, than in animals with more complete lesions which involve both striatal and non-striatal areas (Kirik et al 2001). This indicates that it may be important that part of the non-striatal DA projections remains intact; these spared portions of the host DA system, particularly those innervating non-striatal forebrain areas, may be necessary for the VM grafts to exert their optimal functional effect.

Proof-of-concept in clinical trials

Since 1987 about 350 patients with advanced PD have received transplants of fetal mesencephalic DA neurons (see Lindvall & Hagell 2000, Dunnett et al 2001, Hagell & Brundin 2001). These open-label trials have provided convincing data to show that embryonic human nigral neurons, taken at a stage of development when they have started to express their dopaminergic phenotype, can survive, integrate and function over a long time in the human brain (i.e. in a tissue environment with an ongoing disease process). The overall reduction in UPDRS motor score have been in the range of 30–50%, and positron emission tomography (PET) scans have shown significant increases in [^{18}F]fluorodopa uptake (i.e. dopamine synthesis and retention) in the areas reinnervated by the grafted cells. This increase has been maintained for more than 10 years in some patients. Consistent with the imaging data, good survival of grafted DA neurons and extensive reinnervation of the surrounding host striatum have been demonstrated by immunohistochemistry in two patients that have come to autopsy at 18 months after surgery (Kordower et al 1995, 1996).

Although the functional outcome has been highly variable (see below) long-lasting symptomatic improvement has been observed in about two-thirds of the grafted patients, and in the most successful cases it has been possible to withdraw L-dopa treatment. The majority of patients have shown significant increases in the proportion of time spent in 'on', improvements in the speed and accuracy of movements in defined 'off' (i.e. after drug withdrawal), and maintenance of the improvement with progressive reduction (or complete cessation in several cases) of concurrent L-dopa treatment (see Lindvall & Hagell 2000, Dunnett et al 2001, Hagell & Brundin 2001, for recent reviews).

Lessons from the NIH-sponsored double-blind trials

Based on the encouraging results obtained in the open-label trials the NIH decided in 1993 to provide support for two placebo-controlled double blind trials of fetal VM transplantation in patients with advanced PD. Both trials have now been completed and published. The outcome of these studies has been disappointing, but they are nevertheless important since they provide the first direct evidence of a specific graft-induced improvement, distinguishable from a placebo effect.

In the Denver/Columbia study (Freed et al 2001) the overall improvement in the transplant group ($n=20$) did not reach significance (18% reduction of UPDRS motor score in 'off' at 12 months post-op). In patients younger than 60 years ($n=11$) the improvement of UPDRS at 12 months was 34%. In this trial, the tissue was stored in cell culture for up to 4 weeks before implantation and no immunosuppressive treatment was given. In this trial tissue cultured for up to 4 weeks was used, and less tissue was implanted as compared to the previous open-label trials. In two patients who died after grafting, the number of surviving TH-positive neurons in each putamen was assessed by TH immunohistochemistry at 7000 and 40 000 (Freed et al 2001), which is less than half of what has previously been obtained in functional grafts in the open label trials (ranging from 80 000–135 000 surviving DA neurons; see Kordower et al 1995). These findings underscore the fact that the number of viable DA neurons is a critical factor determining the functional efficacy of intrastriatal VM transplants (Hagell & Brundin 2001).

In the Tampa/Mount Sinai trial (Olanow et al 2003), solid pieces of human fetal mesencephalic tissue from one ($n=11$) or four ($n=12$) donors were implanted in the postcommissural putamen on each side. The transplanted patients were compared to a group of 11 patients subjected to sham-surgery. Immunosuppressive treatment, with cyclosporine only, was given for 6 months after surgery and patients were followed for 2 years. The trial failed to meet its primary outcome, i.e. group difference in UPDRS motor scores at 24 months. Interestingly, however, the patients grafted with tissue from four donors showed the expected progressive improvements up to six months after surgery but deteriorated thereafter, so that by 18 months after transplantation the UPDRS scores were back to preoperative values.

The time-course of functional changes seen in the two double-blind trials is clearly different from that previously observed in the open-label studies. In both trials improvement in UPDRS motor scores was observed during the first 4–6 months after transplantation, but this was not sustained beyond 6 months: in the Denver/Columbia study the improvement seen in the younger patient group levelled off at this time point, while in the Tampa/Mount Sinai study the UPDRS scores gradually returned towards pre-operative values. In the Lund patients (see, e.g. Wenning et al 1997, Piccini et al 2000) many of the transplanted patients have shown continued improvement between 6 and 12 months, and in some patients also during the second postoperative year. Interestingly, the magnitude of symptomatic improvement seen in the Freed et al (2001) and Olanow et al (2003) trials over the first 4–6 months matches fairly well the improvement seen in the Lund patients over this time period, but while the Lund patients continue to improve over the subsequent 12–18 months the patients in the two double-blind studies do not.

This difference is most readily explained by a difference in the immunosuppressive treatments used: in the two NIH-sponsored placebo-controlled trials the immunosuppression was kept to a minimum, i.e. no immunosuppression at all in the Denver/Columbia trial, and a low-dose cyclosporine-only regimen in the Tampa/Mount Sinai study. In contrast, a triple immunosuppressive regimen, with high doses initially, has been maintained for at least 12 months in the Lund patients. Interestingly, in the Tampa/Mount Sinai trial the cyclosporine treatment was stopped after 6 months, i.e. at the time when the functional improvement ceased to increase. It seems highly probable, therefore, that the failure of the transplants to develop maximal functional benefits in these trials is explained, at least in part, by the development of a delayed immunological response to the allogeneic graft tissue, the impact of which became apparent during the second half-year after transplantation.

Earlier animal experimental studies have shown that intracerebral allogeneic transplants which differ immunologically on both major and minor histocompatibility antigens will induce an acute inflammatory response, accompanied by an up-regulation of both class I and class II antigens in the donor cells. This response subsides over the first 6–8 weeks, but in cases where the recipient animal is immunized against the donor tissue the immune response may rebound, leading to a long-lasting inflammatory response coupled to macrophage and microglial activation at the graft site. Hudson et al (1994) and Shinoda et al (1995, 1996) have shown that this delayed inflammatory response may be detrimental to both survival and function of intrastriatal DA neuron transplants. In such chronic immune-activated transplants the DA neurons may survive in a compromised state for a long time, with patches of activated microglia-macrophages and increased expression of class II antigens. Interestingly, this is a picture similar to that observed in the two cases that have come to autopsy in the Tampa/Mount Sinai trial. Solid tissue allografts, as used in this trial, are likely to be more immunogenic since the blood capillaries in such grafts to a large extent are donor-derived and induced to express high levels of donor class I antigens (Sloan et al 1990, Baker-Cairns et al 1996). For this reason solid grafts may be particularly prone to develop this delayed inflammatory response. The immune reaction against the implanted cells will also depend on the extent of tissue damage, or bleeding, at the implantation site(s), and it is further enhanced when the transplantation is made in stages, with several weeks' interval (Duan et al 1993). Long-term, effective immunosuppressive treatment, therefore, is essential to allow allogeneic DA neurons to develop their full functional potential, particularly when the transplantation is performed bilaterally with several weeks interval, or multiple needle penetrations are used.

Ways to improve the overall efficacy of DA neuron transplants

The clinical outcome of fetal DA neuron transplantation has been highly variable. This variability is seen not only between different trials and centres, but also within groups of PD patients transplanted at the same centre. In the Lund program some of the transplanted patients have shown an excellent response, with 40–60% improvement in UPDRS motor score in 'off' at 1–2 years after transplantation, while others have shown little or no benefit. Transplant-induced increases in [^{18}F]fluorodopa uptake have also been highly variable, which suggests that differences in survival and growth of the grafted DA neurons is one important factor in determining the functional outcome in grafted patients. Good functional recovery ($>$30% reduction in UPDRS motor score) is obtained only in patients with at least 40–50% increase in putaminal [^{18}F]fluorodopa uptake in the grafted putamen. Small transplants, poorly surviving transplants, or poor DA fibre outgrowth (less than about 40% increase in overall [^{18}F]fluorodopa uptake in the grafted putamen) have generally been associated with limited graft-induced functional improvement. From the autopsy cases reported by Kordower et al (1995, 1996) it seems that a good functional response is associated with survival of at least 100 000 TH-positive neurons in the putamen and reinnervation of about one-quarter to two-thirds of the putamen volume. This level of graft survival and TH-positive fibre outgrowth may correspond to a recovery of putaminal [^{18}F]fluorodopa uptake to about 50% of normal.

The magnitude of increase in [^{18}F]fluorodopa uptake in the grafted putamen, however, does not always match the level of graft-induced functional improvement. Even in patients that show a marked increase in [^{18}F]fluorodopa uptake, the graft-induced improvement in motor function, i.e. in UPDRS motor score, has been quite variable. This indicates that other factors, such as the selection of patients for DA neuron transplantation, the composition and preparation of the graft tissue, and the placement of the grafts in the brain, are likely to play an important role.

Patient selection

DA cell replacement in PD is based on the assumption that restoration of DA neurotransmission in restricted areas of the forebrain is sufficient to reverse, or significantly ameliorate, motor symptoms. Although degeneration of the nigrostriatal DA pathway is generally viewed as the common unifying defect in idiopathic PD, it is clear from autopsy studies that clinically diagnosed PD patients display a range of other neuropathological changes that are variable in extent and may involve other brainstem and cortical areas as well. Thus, Lewy bodies and α-synuclein-positive inclusions and dystrophic neurites vary markedly in extent and location from patient to patient (Braak & Braak 2000) and widespread cortical changes, including amyloid plaques, neurofibrillary tangles and cell loss, are frequently observed, also in patients without clinically diagnosed dementia (Perl et al 1998).

Some of these changes may reflect the progressive nature of the disease, such that areas outside the nigrostriatal system may become more severely involved as the disease progresses. If so, DA neuron transplants would be expected to be more efficacious during early stages of PD, i.e. at stages when the pathological changes are likely to be confined mostly to the nigrostriatal DA pathway. This may, at least in part, explain why the overall improvement seen in the Tampa/Mount Sinai trial was so poor. As judged by the disease rating scores and the doses of antiparkinsonian medication taken by the patients at the time of surgery, these patients are likely to have had overall more advanced disease than, for example, the patients in the Lund program. Freed et al (2001) have reported that transplant-induced recovery was significantly better in younger patients. In the older patient group (> 60 years of age) the reduction in UPDRS motor score seen after transplantation was significantly correlated to the magnitude of the response to L-dopa medication as assessed pre-operatively (C. R. Freed, personal communication).

In support of this, rodent studies have shown that the survival and growth of fetal mesencephalic DA neurons are much reduced (by about 75%) in aged recipients, and that the functional effect of identical intrastriatal DA neuron transplants is less in older animals (Sortwell et al 2001). This age-related decline in the survival and function of fetal VM grafts have been proposed to be due to a reduced production of neurotrophic factors necessary for DA neuron growth and survival in the aged host striatum (Carvey et al 1996, Ling et al 2000, Yurek & Fletcher-Turner 2001, 2002).

Available clinical and experimental data thus point to both age and disease severity as important factors in determining functional efficacy, and hence also the variable outcome, of intrastriatal VM transplants. Moreover, the impact of these two factors may be additive: advancing age is likely to affect above all the ability of the denervated striatum to sustain survival and growth of the transplanted DA neurons. Increased disease severity, on the other hand, will affect the ability of the host brain to respond to otherwise fully functional DA neuron grafts. From what we know today it seems likely that DA neuron transplants will be optimally effective, and hence therapeutically valuable, only in moderately advanced PD patients exhibiting a good and reliable response to L-dopa medication in combination with a symptomatology and changes on [^{18}F]fluorodopa PET scans that suggest nigrostriatal dysfunction as the primary and leading cause of disability.

Composition and preparation of the graft tissue

There has so far been no attempt to standardize the way fetal cell transplantation has been carried out at different centres. Almost all aspects of tissue procurement and handling varies from one center to another: the dissection of the tissue, the age of the donor fetuses, the length and type of storage after dissection, the way the tissue is dissociated prior to grafting (into pieces or crude cell suspensions), and the composition of the medium used for storage and/or injection. As a result, the composition of the cell material used for transplantation is likely to vary significantly. In the Denver/Columbia trial the tissue was cultured for up to 4 weeks prior to grafting, and in some cases a growth factor cocktail was used. The tissue was implanted as non-dissociated solid tissue strands. In the Tampa/Mount Sinai trial tissue was stored in so-called hibernation medium at 4 °C for up to 2 days and the grafts were implanted as solid pieces. In the Lund programme we used non-stored tissue in most cases (within about 5–6 hours after dissection) and the tissue has been dissociated into a crude cell suspension prior to injection.

The use of solid tissue grafts is partly justified by the fact that this makes it possible to use somewhat older aborted fetuses (up to 9–9.5 weeks of gestation compared to 6–8 weeks for dissociated tissue). The negative aspect of using solid tissue grafts is that they are likely to be more immunogenic. Experiments with intracerebral allografts in rodents have shown that the blood capillaries that develop in grafts composed of solid tissue pieces are almost entirely of donor origin and will express high levels of class I HLA antigens for weeks or months after transplantation, while the capillaries in suspension grafts are mostly of host origin (Sloan et al 1990, Baker-Cairns et al 1996). The disadvantage of using long-term stored or cultured tissue is that the cellular composition is likely to change significantly over time in culture. Cultured VM DA neurons, moreover, survive less well after transplantation and show more limited fibre outgrowth than freshly dissociated DA neuroblasts (Brundin et al 1988).

Graft placement

In the clinical trials performed so far, the VM grafts have—with few exceptions—been placed in the putamen, and in some cases only in the posterior putamen. The outgrowing axons can be estimated to extend 2–3 mm from the site of implantation, which means that the functional effect of the transplanted DA neurons will be limited to a relatively restricted area surrounding the cell deposits. We have previously estimated that about a third to a half of the grafted putamen will be reached by the graft-derived DA innervation from three or four implantation sites. Remaining parts of the putamen, as well as other striatal areas, are left untouched. The reason for selecting the putamen as the primary transplantation target is that this region is often the one exhibiting the most marked reductions in DA content, and this region is also the part of the striatal complex that is physiologically most closely linked to motor control. However, it is well known that the loss of DA is quite widespread and involves other parts of the basal ganglia and the limbic forebrain, as well as areas of the cortex. In most of these regions the mean reductions, as measured biochemically in post-mortem brain samples from PD patients, is in the range of 50–90% (see Agid et al 1987).

The extent to which areas outside the putamen are involved (e.g. caudate nucleus, nucleus accumbens and frontal cortex) varies from patient to patient and may also be reflected in the severity and range of symptoms exhibited by each patient. Based on these considerations it seems likely that the placement of VM transplants has to be tailored to each patient. The way to approach this, we believe, is to perform high-resolution [^{18}F]fluorodopa PET scans prior to surgery in order to identify the areas, or subregions, where [^{18}F]fluorodopa is most severely reduced. Such data will be useful not only as a rational basis for selecting optimal transplantation sites for each patient, but they will also help in establishing a correct diagnosis and linking specific symptoms to the pattern of striatal DA dysfunction.

The striatum and associated limbic and cortical forebrain areas are known to be functionally heterogeneous. As a consequence, the effect of DA denervation on sensorimotor behaviour is directly dependent on which area, or areas, are affected by the denervated lesion. Although the nigrostriatal DA pathway may be the one most severely affected in PD, there is clearly a variable involvement of the meso-cortico-limbic DA pathway as well. This raises the question as to whether grafts limited to the caudate nucleus and putamen may be sufficient in all cases. Experiments in 6-OHDA-lesioned rats have shown that the magnitude of functional recovery induced by intrastriatal VM transplants is greater in partially lesioned animals where the limbic and cortical projections are spared, and that the effect of identical VM transplants is less in animals with complete lesions (Kirik et al 2001). This indicates that spared portions of the DA projection system innervating non-striatal areas may be necessary for the intrastriatal VM grafts to exert their optimal functional effect. The implication of this is that patients with more widespread DA neuron cell loss, where the disease has progressed to involve also non-striatal areas, may be less suitable candidates for intrastriatal VM transplants.

Generation of DA neurons from embryonic and neural stem cells

The problems associated with the use of primary human fetal tissue make it necessary to find alternative sources of cells for intracerebral grafting. Ideally, these cells should have the potential for expansion in large numbers *in vitro* and allow differentiation into fully functional dopaminergic neurons upon transplantation and/or after induction *in vitro*. Expandable neural progenitors may be derived from several sources: from ES cells, or from fetal or adult CNS tissue. In principle, there are two ways of using such progenitors for transplantation in PD: the cells can either be pre-differentiated *in vitro* into DA neurons before grafting, or a committed precursor cell type may be identified that can differentiate into an appropriate mesencephalic dopaminergic neuronal phenotype after transplantation. All attempts at using neural progenitors of human origin have so far failed to generate any significant number of dopaminergic neurons from this source in culture, and it has proved even more difficult to obtain any surviving, functional dopaminergic neurons *in vivo* after transplantation of human neurosphere cells to the striatum (see Svendsen & Caldwell 2000, Lindvall 2003, for discussion).

Some so-called TH-induction protocols have been established for *in vitro*-expanded rodent progenitors. For example, Studer et al (1998, 2000) have published a two-step procedure for expansion of rat mesencephalic precursors in the presence of bFGF (basic fibroblast growth factor; FGF2), followed by differentiation into TH-positive neurons upon removal of the mitogen. These cells were shown to survive and exhibit some functional impact after transplantation into the rat model of PD, although the survival rate of the expanded and pre-differentiated cells was very low. Other investigators have used a mixture of cytokines, mesencephalic membrane fragments and striatal conditioned media resulting in stable DA differentiation of rat fetal mesencephalic precursors (Ling et al 1998, Potter et al 1999). Interestingly, these cells survived transplantation to the striatum in 6-OHDA lesioned rats and provided some functional benefit although the survival rate of the grafted cells was quite low (Carvey et al 2001).

The most promising results reported so far have been obtained with mouse embryonic stem (ES) cells. These cells can differentiate into TH-positive, DA-producing neurons either spontaneously (Björklund et al 2002), or after pre-differentiation (Lee et al 2000). These ES cell-derived cells survive very well after transplantation to the striatum, and give significant restoration of function after grafting into the rat 6-OHDA lesion model (Björklund et al 2002, Kim et al 2002). Further studies from Isacson's and McKay's labs have shown that the proportion of the ES cells that differentiate into TH-positive neurons can be further promoted by induced expression of Nurr1, a transcription factor of the retinoic acid nuclear receptor superfamily which is required for the normal development of midbrain DA neurons (Kim et al 2002, Chung et al 2002).

Although these ES cell-derived TH-positive neurons clearly are DA producing and express, at least in part, a mesencephalic neuronal phenotype, it is unclear to what extent they differentiate into fully functional DA neurons of the correct type, i.e. DA neurons of the type normally present in the substantia nigra. The DA neurons located in the ventral mesencephalon are of two distinct subtypes: the A9 neurons of the substantia nigra and the A10 neurons of the ventral tegmental area (VTA). Transplants of fetal VM tissue contain a mixture of both subtypes. Previous experiments using fetal VM transplants (Schultzberg et al 1984, Haque et al 1997, Isacson et al 2003) suggest that the transplanted DA neurons that reinnervate the host striatum are of the A9 subtype (which is the one normally innervating the caudate nucleus and putamen). The fibre outgrowth from the A10 neurons, by contrast, seems to be confined to the graft itself and is thus unlikely to exert any major functional effect. There are good reasons to believe, therefore, that the functional impact of intrastriatal VM grafts is predominantly mediated by the A9 neuron subtype. Indeed, the morphology of the TH-positive cells in intrastriatal ES cell transplants, and the relatively restricted TH-positive fibre outgrowth from these grafts into the host striatum, suggest that only a small fraction of the ES cell-derived DA neurons are of the A9 type. This is important to keep in mind when evaluating the outcome of DA cell transplants in either animals or patients; the total number of surviving TH-positive neurons in the graft does not tell us how many of these are fully functional A9 neurons with projections extending into the host striatum.

Conclusions

The usefulness of *in vitro* generated DA neurons for transplantation in PD will depend not only on their ability to adopt a fully functional dopaminergic fate, but also on their capacity to establish appropriate functional synaptic connections with striatal target neurons. It is important to keep in mind that supportive glial cells, and also other types of neurons, may contribute to survival, differentiation and growth, as well as the functional competence of transplanted DA neurons. If so, an enriched population of pre-differentiated DA neurons may not be the optimal cell preparation. Although stem cell technology holds promise to turn cell therapy from a highly experimental procedure into clinically useful routine therapies, the complexity of the biological problems should not be underestimated. The mechanisms involved in the regulation of DA neuron differentiation, connectivity and function, and the role of the host tissue response, immune/inflammatory reactions, and supporting glial elements in these processes, clearly need to be explored in much greater detail. Progress along these lines should be made with great care; premature clinical trials of stem cell based procedures in the absence of extensive and convincing pre-clinical safety and efficacy studies may seriously hinder further progress in this field.

References

Agid Y, Javoy-Agid F, Ruberg M 1987 Biochemistry of neurotransmitter in PD. In: Marsden CD, Fahn S (eds) Movement disorders 2. Butterworth, London, p 166–230

Baker-Cairns BJ, Sloan DJ, Broadwell RD, Puklavec M, Charlton HM 1996 Contributions of donor and host blood vessels in CNS allografts. Exp Neurol 142:36–46

Björklund A 1992 Dopaminergic transplants in experimental parkinsonism: cellular mechanisms of graft-induced functional recovery. Curr Opin Neurobiol 2:683–689

Björklund LM, Sanchez-Pernaute R, Chung S et al 2002 Embryonic stem cells develop into functional dopaminergic neurons after transplantation in a Parkinson rat model. Proc Natl Acad Sci USA 99:2344–2349

Braak H, Braak E 2000 Pathoanatomy of Parkinson's disease. J Neurol 247(suppl 2):II3–10

Brundin P, Hagell P 2002 The neurobiology of cell transplantation in Parkinson's disease. Clin Neurosci Res 1:507–520

Brundin P, Barbin G, Strecker RE, Isacson O, Prochiantz A, Björklund A 1988 Survival and function of dissociated rat dopamine neurones grafted at different developmental stages or after being cultured in vitro. Brain Res 467:233–243

Carvey PM, Lin DH, Faselis CJ, Notermann JK, Ling ZD 1996 Loss of striatal DA innervation increases striatal trophic activity directed at DA neurons in culture. Exp Neurol 140:184–197

Chung S, Sonntag KC, Andersson T et al 2002 Genetic engineering of mouse embryonic stem cells by Nurr1 enhances differentiation and maturation into dopaminergic neurons. Eur J Neurosci 16:1829–1838

Duan W-M, Widner H, Björklund A, Brundin P 1993 Sequential intrastriatal grafting of allogeneic embryonic dopamine-rich neuronal tissue in adult rats: will the second graft be rejected? Neuroscience 57:261–274

Dunnett SB, Björklund A, Lindvall O 2001 Cell therapy in Parkinson's disease — stop or go? Nat Rev Neurosci 2:365–369

Freed CR, Greene PE, Breeze RE et al 2001 Transplantation of embryonic dopamine neurons for severe Parkinson's disease. New Engl J Med 344:710–719

Hagell P, Brundin P 2001 Cell survival and clinical outcome following intrastriatal transplantation in Parkinson disease. J Neuropathol Exp Neurol 60:741–752

Haque NS, LeBlanc CJ, Isacson O 1997 Differential dissection of the rat E16 ventral mesencephalon and survival and reinnervation of the 6-OHDA-lesioned striatum by a subset of aldehyde dehydrogenase-positive TH neurons. Cell Transplant 6:239–248

Hudson JL, Hoffman A, Stromberg I, Hoffer BJ, Moorhead JW 1994 Allogeneic grafts of fetal dopamine neurons: behavioral indices of immunological interactions. Neurosci Lett 171:32–36

Isacson O, Bjorklund LM, Schumacher MD 2003 Toward full restoration of synaptic and terminal function of the dopaminergic system in Parkinson's disease by stem cells. Ann Neurol 53(suppl 3):S135–S148

Kim JH, Auerbach JM, Rodriguez-Gomez JA et al 2002 Dopamine neurons derived from embryonic stem cells function in an animal model of Parkinson's disease. Nature 418:50–56

Kirik D, Winkler C, Björklund A 2001 Growth and functional efficacy of intrastriatal nigral transplants depend on the extent of nigrostriatal degeneration. J Neurosci 21:2889–2896

Kordower JH, Freeman TB, Snow BJ et al 1995 Neuropathological evidence of graft survival and striatal reinnervation after the transplantation of fetal mesencephalic tissue in a patient with Parkinson's disease. N Engl J Med 332:1118–1124

Kordower JH, Rosenstein JM, Collier TJ et al 1996 Functional fetal nigral grafts in a patient with Parkinson's disease: chemoanatomic, ultrastructural, and metabolic studies. J Comp Neurol 370:203–230

Lee SH, Lumelsky N, Studer L, Auerbach JM, McKay RD 2000 Efficient generation of midbrain and hindbrain neurons from mouse embryonic stem cells. Nat Biotechnol 18:675–679

Lindvall O 2003 Stem cells for cell therapy in Parkinson's disease. Pharmacol Res 47:279–287

Lindvall O, Hagell P 2000 Clinical observations after neural transplantation in Parkinson's disease. Prog Brain Res 127:299–320

Ling ZD, Potter ED, Lipton JW, Carvey PM 1998 Differentiation of mesencephalic progenitor cells into dopaminergic neurons by cytokines. Exp Neurol 149:411–423

Ling ZD, Collier TJ, Sortwell CE et al 2000 Striatal trophic activity is reduced in the aged rat brain. Brain Res 856:301–309

Nikkhah G, Cunningham MG, Jödicke A, Knappe U, Björklund A 1994 Improved graft survival and striatal reinnervation by microtransplantation of fetal nigral cell suspensions in the rat Parkinson model. Brain Res 633:133–143

Olanow CW, Goetz CG, Kordower JH et al 2003 A double-blind controlled trial of bilateral fetal nigral transplantation in Parkinson's disease. Ann Neurol 54:403–414

Perl DP, Olanow CW, Calne D 1998 Alzheimer's disease and Parkinson's disease: distinct entities or extremes of a spectrum of neurodegeneration? Ann Neurol 44 (suppl 1):S19–S31

Piccini P, Lindvall O, Bjorklund A et al 2000 Delayed recovery of movement-related cortical function in Parkinson's disease after striatal dopaminergic grafts. Ann Neurol 48:689–695

Potter ED, Ling ZD, Carvey PM 1999 Cytokine-induced conversion of mesencephalic-derived progenitor cells into dopamine neurons. Cell Tissue Res 296:235–246

Schultzberg M, Dunnett SB, Björklund A et al 1984 Dopamine and cholecystokinin immunoreactive neurones in mesencephalic grafts reinnervating the neostriatum: evidence for selective growth regulation. Neuroscience 12:17–32

Shinoda M, Hudson JL, Stromberg I, Hoffer BJ, Moorhead JW, Olson L 1995 Allogeneic grafts of fetal dopamine neurons: immunological reactions following active and adoptive immunizations. Brain Res 680:180–195

Shinoda M, Hudson JL, Stromberg I, Hoffer BJ, Moorhead JW, Olson L 1996 Microglial cell responses to fetal ventral mesencephalic tissue grafting and to active and adoptive immunizations. Exp Neurol 141:173–180

Sloan DJ, Baker BJ, Puklavec M, Charlton HM 1990 The effect of site of transplantation and histocompatibility differences on the survival of neural tissue transplanted to the CNS of defined inbred rat strains. Prog Brain Res 82:141–152

Sortwell CE, Camargo MD, Pitzer MR, Gyawali S, Collier TJ 2001 Diminished survival of mesencephalic dopamine neurons grafted into aged hosts occurs during the immediate postgrafting interval. Exp Neurol 169:23–29

Studer L, Tabar V, McKay RD 1998 Transplantation of expanded mesencephalic precursors leads to recovery in parkinsonian rats. Nat Neurosci 1:290–295

Studer L, Csete M, Lee SH et al 2000 Enhanced proliferation, survival, and dopaminergic differentiation of CNS precursors in lowered oxygen. J Neurosci 20:7377–7383

Svendsen CN, Caldwell MA 2000 Stem cells in the developing central nervous system: implications for cell therapy through transplantation. Prog Brain Res 127:13–34

Wenning GK, Odin P, Morrish P et al 1997 Short- and long-term survival and function of unilateral intrastriatal dopaminergic grafts in Parkinson's disease. Ann Neurol 42:95–107

Winkler C, Bentlage C, Nikkhah G, Samii M, Björklund A 1999 Intranigral transplants of Gaba-rich Striatal tissue induce behavioral recovery in the rat Parkinson model and promote the effects obtained by intrastriatal dopaminergic transplants. Exp Neurol 155:165–186

Winkler C, Kirik D, Björklund A, Dunnett SB 2000 Transplantation in the rat model of Parkinson's disease: ectopic versus homotopic graft placement. Prog Brain Res 127:233–265

Yurek DM, Fletcher-Turner A 2001 Differential expression of GDNF, BDNF, and NT-3 in the aging nigrostriatal system following a neurotoxic lesion. Brain Res 891:228–235

Yurek DM, Fletcher-Turner A 2002 Temporal changes in the neurotrophic environment of the denervated striatum as determined by the survival and outgrowth of grafted fetal dopamine neurons. Brain Res 931:126–134

DISCUSSION

Nadal-Ginard: How many patients have received fetal cell transplants?

Björklund: In the open-label studies it is around 350 patients that have been reported in one form or another. These reports are of variable quality.

Weissman: I'd like to address the decision in both the experimental and clinical models to put the cells directly in the striatum. The normal position of the cells is in the substantia nigra, and except for perhaps one report there is very little evidence of turnover of cells in this region. What do we know about the initial axon guidance during development if you were to put the cells in the substantia nigra?

Björklund: We have transplanted fetal dopamine neurons to the substantia nigra in rats at different ages (Bentlage et al 1999). In the early neonatal period the grafted dopamine neuroblasts can extend axons to the striatum. This is up to about 10 d of age. Between 10 and 20 d there are changes in the properties of the territory so that by 20 d it looks like an adult. If we place the dopamine neurons in the substantia nigra, or somewhere along the nigro–striatal pathway, in adult rats the grafted neurons show no tendency to grow axons along the nigro–striatal pathway towards the striatum. This means that the outgrowing axons remain confined to the vicinity of the grafted cells. Nevertheless, dopamine neurons grafted to the substantia nigra do exert some functional effects. There are quite a few studies looking at dopamine neuron transplants in the nigra, but they are less functional than those placed in the striatum, i.e. close to the dopamine responsive striatal neurons.

Weissman: Do we know much about axon guidance in this particular developmental place compared with other ones in the spinal cord?

Björklund: We know there are several kinds of inhibitory molecules, some of which are associated with myelin. None of these have been studied properly in the nigrostriatal pathway. The only study that has a bearing on this is from James Fawcett's lab (Moon et al 2001) in which they used injection of the chondroitinase enzyme to block the formation of inhibitory chondroitin sulfate proteoglycans. They reported an improved axon growth along the nigrostriatal pathway after this treatment.

Reference

Bentlage C, Nikkah G, Cunningham MG, Björklund A 1999 Reformation of the nigrostriatal pathway by fetal dopaminergic micrografts into the substantia nigra is critically dependent on the age of the host. Exp Neurol 159:177–190

Moon LD, Asher RA, Rhodes KE, Fawcett JW 2001 Regeneration of CNS axons back to their target following treatment of adult rat brain with chondroitinase ABC. Nat Neurosci 4:465–466

Ethical (and political) issues in research with human stem cells

Thomas H. Murray

The Hastings Center, 21 Malcolm Gordon Road, Garrison, NY 10524-5555, USA

Abstract. Though it is possible to discuss the ethics of research with human embryonic stem (ES) cells without any reference to politics, doing so would shed little light on the nature or intensity of the public debate, especially in the USA. Ethical arguments and scientific claims have been appropriated and offered with what appears often to be little regard for their soundness or relevance. In this paper I will discuss: (1) briefly, the history of research ethics and policy regarding human embryos in the US and the UK; (2) the decisions made by the current President of the USA regarding ES cell research, including a critical examination of the ethical arguments offered; (3) the evolution of the debate in which nuclear transplantation in stem cell research has been conflated with cloning to attempt to create a human infant; (4) the stalemate within the US Congress over legislation on cloning; (5) possible developments suggesting that the current US policy on ES cell research may become unstable; and (6) the broader context of the ethical debate and what is at stake for the major protagonists.

2005 Stem cells: nuclear reprogramming and therapeutic applications. Wiley, Chichester (Novartis Foundation Symposium 265) p 188–203

The irony did not escape me: I was explaining to the US Senate Judiciary Committee and its Chairman, Senator Orrin Hatch, how the arguments against research cloning in human stem cells offered by Leon Kass, Chair of President Bush's Council on Bioethics, were different and more subtle than the broadsides against all research with human embryos fired by right-to-life advocates. It was clear to me that I had been invited to the hearings as a counterweight to Dr Kass; and I had filled that role, criticizing what I saw as misrepresentations of science, laying out the ethical case for why research on human stem cells was justifiable, and warning about the conflation of nuclear transfer in stem cell research with human reproductive cloning. Yet here I was defending Kass against having his own views — with which I disagree — misunderstood (Committee on the Judiciary 2003).

In the controversy over research with human stem cells, science and ethics have been wielded as political weapons. The fault is not entirely one-sided. Scientists defending human stem cell research have fed hopes of wonderful new therapies just around the corner. Opponents have repeatedly misrepresented the virtues of other sources of stem cells — from adults, cord blood — as equivalent or superior to embryonic stem (ES) cells. (Senator Sam Brownback, a leader of the opposition to ES cell research, testified at the same hearing, brandishing a thick binder containing, he claimed, an abundance of evidence demonstrating the therapeutic utility of other stem cell sources. Senators Hatch and Brownback engaged in a rhetorical dance, Hatch inviting him to submit the documents in the binder as evidence and offering to have eminent scientists evaluate it, Brownback dodging the invitation, never relinquishing his grip on the binder. I was told later that it relied heavily on anecdotal reports of dubious scientific merit.) These manoeuvres might appear amusing if not for the stakes: the hearing was to discuss a bill that would have criminalized all nuclear transfer into human embryos in the USA, even for stem cell research (Public Health Service Act 2003). Scientists violating this law could be imprisoned for ten years and fined a million dollars or more. Senator Hatch, a staunch opponent of abortion, nonetheless found within his Mormon tradition reasons to distinguish stem cell research, even nuclear transfer in stem cells, from abortion. Hatch was determined not to allow Brownback's bill to pass.

Just as has happened with the science of stem cells, discourse on the ethics of research with human ES cells has been exploited for political advantage. (I am relieved, however, to report that no magical binder of moral arguments has yet been flourished before a Congressional committee — at least not to my knowledge.) Ethical arguments are, of course, at the heart of this debate. But to ignore their interactions with the politics of the debate is to be wilfully naïve. Their political currency does not make ethical arguments stronger or weaker; but it does help to explain why certain ethical considerations rise to the fore, while others quietly sink beneath the public's horizon. I will try to explain how certain considerations and arguments persist and come to dominate US public policy on human ES cells. And I will challenge the scientific community to respond to the ethical and policy challenges.

The USA and the UK present an intriguing contrast in their respective public policies concerning human embryos — in research and in therapeutic contexts such as IVF and other assisted reproductive technologies. From the late 1970s American public policy permitted funding for research on human embryos, if and only if the proposed research was approved by a body known as the Ethics Advisory Board (EAB), which was appointed by a member of the President's Cabinet, the Secretary of the Department of Health, Education and Welfare. That Department's functions have been reshuffled, with the health and research related responsibilities passed down to the Department of Health and Human Services (DHHS). By 1979, in response to a research proposal involving IVF, the EAB concluded that federal funding for such research was ethically acceptable if it met certain conditions including consent from the gamete providers, an important scientific goal not reachable by other approaches, and a prohibition on maintaining an embryo beyond the stage reached after 14 days of embryonic development. The Secretary never acted upon the EAB's report; the EAB was disbanded in 1980 (National Bioethics Advisory Commission 1999).

From 1980 until 1993 American public policy on embryo research presented a Catch 22: embryo research could be funded by the federal government if it was approved by a committee that did not exist. Shortly after the inauguration of President Clinton in 1993 the Catch 22 was removed, but federal funding for research on human embryos was delayed until a Human Embryo Research Panel, convened by the Director of the National Institutes of Health, made its recommendations. This panel proposed funding research on human embryos, hedged in with many safeguards and protections. In its most controversial action the panel concluded that under very rare, carefully circumscribed conditions the federal government should also fund the creation of human embryos for research (National Institutes of Health Human Embryo Research Panel 1994). Even before the panel's report was officially issued, the President declared that there would be *no* federal support for creating research embryos (Office of the White House Press Secretary 1994) and the US Congress then effectively prohibited federal funding for all research on embryos.

In 1999 a legal opinion issued by the General Counsel of DHHS distinguished between the embryos that were the source of ES cells, and the ES cell lines themselves. The opinion concluded that although Congress had prohibited federal funding for work on embryos, that prohibition did not apply to cell lines derived from those embryos; NIH could, therefore, fund research on ES cell lines (National institutes of Health Office of the Director 1999). On 9 August 2001 President Bush ruled that federal research funds could only be used for ES cell lines derived *before* the speech in which he announced and defended his policy (Bush 2001). Ever since, debate in the USA has raged over the ethical justification given for and the scientific constraints imposed by this policy.

The inconsistencies within the borders of the USA are stunning. All of the policies above pertained only to *federal* support of research on human embryos and, later, ES cells. On the research side anyone who uses private funds and can effectively sequester their research on embryos from any federal support they may receive can do virtually anything they want with human embryos. There are some caveats: some states prohibit or restrict embryo research; if the researcher works at an academic institution that requires IRB approval for all studies, not merely federally funded ones, the researcher may have to persuade the IRB that the research was ethically sound. But scientists who work at privately funded independent laboratories (such as Advanced Cell Technology or Geron) can work with few constraints. Furthermore, IVF clinics create, discard, freeze or implant thousands of embryos each year. A recent survey of IVF clinics ascertained that nearly 400 000 embryos were stored in the USA. Some of the largest clinics declined to participate in that survey, so the number is surely an underestimate. 11 283 stored embryos were offered for research by the couples responsible for them, more than the number designated for donation to another couple (9225) or for destruction (8840) (Hoffman et al 2003).

The Hastings Center issued a report in 2003 on Reprogenetics calling for more public accountability and control over practices in IVF clinics. Our report cited the UK's HFEA as an example of how a nation with strong cultural affinities to the USA created a structure to respond to research and clinical uses of human embryos (Parens & Knowles 2003).

We need to turn now from policy and politics to ethics. The ethical arguments in favour of not criminalizing nuclear transfer in human stem cells are straightforward. The most compelling reason is that this research may contribute in time to the relief of suffering and the postponement of untimely death. Success is of course not certain; it is also possible that the greatest contributions to human health from research cloning will come from the basic research it will make possible as scientists create stem cell lines for an enormous variety of diseases, cell lines that will allow us to understand and, ultimately, treat or prevent those diseases. Some critics of ES cell research claim that the impetus for it grows out of selfish desires to cling greedily to life. What is sometimes overlooked is the deep human truth that suffering and death afflicts families, not merely the individual. Our lives are entwined with the lives of others whom we love. Their suffering, their death, profoundly affects our own. When we minister to suffering, we minister not only to the individual but also to all of those who love and care for her or him. Anyone of us who has loved someone who suffered or died knows the truth of this.

A second argument appeals to our moral, legal and political traditions of freedom of speech and freedom of inquiry. Americans value the quest for new frontiers; today's explorers are more likely to wear white coats and inhabit laboratories than to paddle canoes or hike over mountain passes. Scientific inquiry is obliged to respect moral limits. That principle was resoundingly affirmed in the trials at Nuremberg and in our own nation's apology to the subjects of the Tuskegee syphilis study. But when we have no consensus that a form of research is ethically improper, the wiser course is to allow people to follow their individual consciences. This respects the value of freedom of inquiry without forcing individuals to violate their beliefs.

What reasons do people give for criminalizing nuclear transfer to create stem cells? This is a very important question: it is one thing to decide not to fund an activity because some Americans have moral objections to it. (If we applied that principle broadly, there would be no public funding of research on blood transfusion or, for that matter, transfusions themselves, on the grounds that Jehovah's Witnesses object to transfusions. The same would be true of all research using animals given the opposition of a passionate minority.) It is quite another to create a new federal crime for doing what many Americans do not find inherently wrong — that is, work with very early human blastocysts.

The arguments against permitting nuclear transfer for stem cell research fall into two rough categories: arguments that do not depend upon a belief that embryos hold the same moral status as children and adults, and arguments that require that premise as a starting point. The arguments offered by Kass and those with similar beliefs are in the first category.

In his 2003 testimony before the US Senate Judiciary Committee, Kass made three broad arguments against permitting what he asserts should be called cloning-for-research. (A fourth objection is to the details of the alternative bill supported by Hatch and a majority of Senators.) The first broad argument is that banning reproductive cloning but permitting cloning-for-research is ineffective and counterproductive. Kass envisions industrial-scale production of human embryos, with doctors secretly implanting cloned embryos in women's uteruses—a process he described in an earlier article as 'simple enough that its final steps could be self-administered by the woman...' (Kass 2001). Cloned pregnancies would go undetected and, even if one was uncovered, no state would compel the woman to abort her cloned fetus.

The prospect of mass embryo production is, to say the least, implausible, if for no other reason than the scarcity of human oocytes. It is true that some renegade physician might find a woman willing to try to carry a cloned embryo. But Kass' comment about enforcement misses the mark: enforcement should be against the physician and his or her collaborators, not against the woman. In the extremely unlikely event that the cloned embryo would make it to birth, this would not be the end of civilization. It would be an occasion for immense sadness, most of all for the cloned child, but not remotely comparable to a terrorist detonating a nuclear weapon in the heart of London or New York. The world would survive.

Kass does make an important point: that what we learn from performing nuclear transfer in cloning-for-research would be appropriated by people eager to do reproductive cloning. This is a palpable concern.

The second important argument Kass advances is that cloning-for-research violates important moral boundaries and pushes us headlong down a slippery slope. Deliberately producing embryos for research and then insisting on their destruction, he claims, sets us on the path of 'ever-expanding genetic mastery of one generation over the next' (Committee on the Judiciary 2003). So called 'slippery slope' arguments are commonplace in ethics. But the success of any particular argument rests on two elements:

- whether the empirical assumptions are plausible — that doing X will lead inexorably to Y; and
- whether the bottom of the hill is to be feared and avoided.

I agree with Kass that dangers lurk at the bottom of this particular slope (see my '*The worth of a child*'; Murray 1996). But I am not at all convinced that this particular step onto the hillside will plunge us unavoidably into the precipice. Life is lived on slippery slopes. It is the task of ethics, law and social institutions and practices to carve out solid level ground where we can stand firm and safe. Should we regard ourselves (and our descendants) as the helplessly passive victims of technology, or should we burden ourselves with the task of locating and carving out a firm footing?

The third major claim Kass argues for is that research involving nuclear transfer in stem cells is scientifically and clinical unnecessary. In his Senate Judiciary Committee testimony he said: 'The putative benefits of cloning research are at best speculative, and it is unlikely to be the solution for the immune rejection problem.' He refers to the prospect of overcoming rejection with stem cell therapies derived from transferring the patient's own nucleus as 'not likely to succeed as an effective or practical form of regenerative medicine. Its alleged promise is vastly overrated, not to say spurious.' He offers research on 'multipotent cells' from existing persons and research on animal embryos as adequate substitutes for both basic research and potential clinical applications.

There would be significant ethical, practical, scientific, and regulatory hurdles to be cleared before somatic cell nuclear transfer (SCNT) in stem cells could become a source of useful and affordable therapies. Kass is correct to point this out. But it is not so clear that research on adult stem cells and animal embryos and stem cells is a fully equivalent or adequate substitute. The facts matter here: I suspect the greatest distortions of the scientific facts pertain to this particular question with opponents to ES cell research exaggerating the evidence in favour of stem cells of other origins. We do not know which form of stem cells holds the greatest scientific or clinical promise, and we should not pretend that we do, nor should we base our public policies on misrepresentations of current science, magic binders notwithstanding.

I said there is a second major objection to research on ES cells premised on the belief that embryos hold the same moral status as children and adults. We must acknowledge that morally thoughtful people are not of one mind on the moral status of four or six day-old blastocysts. In *The worth of a child* I posed a challenge: imagine some new ethical argument or scientific fact that persuaded nearly everyone on one side of the embryo-as-person debate that they had been mistaken. The other side is right, they would admit. Can you imagine such an argument or fact? I cannot. Notice that I did not say which side came up with the persuasive new moral consideration. This is, I believe, not because people are impervious to logic, but because our beliefs about embryos are woven into a complex tapestry of other beliefs — about what it means to be a woman, a man, a child; about families; about the importance of being a nurturing parent to the flourishing of adults — men as well as women. This tapestry of beliefs and commitments affects everything from our attitudes towards sex discrimination in employment to the importance of family leave and educational opportunities for women. It also grounds our beliefs about embryos.

Respecting the diversity of beliefs about families, about women, men, children — and embryos — honours our most noble traditions. Where there is a clear and ringing consensus, as there is against cloning to create a child, let us act on it. Where there is profound and principled disagreement, let our laws respect that. Declining to fund such research can be an honourable choice and a defensible public policy. Sending scientists to prison for 10 years and subjecting them to fines of a million dollars or more devalues the ethical views of Americans for whom the possibility of alleviating suffering justifies research cloning.

There are important positive steps we can take now to control destructive uses of the technology. We can insist that all such research, whether publicly or privately funded, must be conducted according to the most stringent ethical standards. We can enhance public accountability of the infertility industry by establishing stringent standards for procuring human eggs — now left, in the USA at least, to an essentially unregulated market. Let us take those steps now.

What, specifically, can the scientific community do?

It can protest when science is misrepresented and misused for political ends. To retain its credibility, the scientific community must insist upon truth and accuracy *even when* the misrepresentations or exaggerations are intended to benefit science. Those short-term benefits damage the public's trust of science in the long run.

Scientists can deepen their understanding of the ethical dispute over stem cells and cloning. This is *not* a conflict between science and religion. There are religious arguments in favour of stem cell research and secular arguments against it, not just the other way round. But the conflict does reflect fundamental moral convictions about what good and fulfilling lives are for women and for men, and whether the paths to fulfilment are radically separate, or mostly shared. Don't expect any new scientific insight or empirical fact about human embryos to convince right-to-life advocates that they have been mistaken in their moral beliefs, and don't think them irrationally stubborn if they don't bow immediately to your brilliant new argument. Moral beliefs that are tied into questions of identity and purpose are woven into tapestries that take a lifetime to create. Altering a central thread is possible but immensely difficult; it requires elaborate and far reaching reconstruction of the entire fabric.

Embryo politics and culture wars will not yield to frontal assault. Scientists can, however, proceed vigorously with research likely to lead to promising new therapies for diseases that cause widespread suffering and early death. The emergence of such therapies may pry open the latent cracks within the right-to-life movement over the moral significance of the earliest *in vitro* embryos. It was remarkable how rapidly the early violent opposition to the President's stem cell research policy within the right-to-life movement went silent. Why? Because, perhaps, the movement recognized disagreements within its ranks over the moral significance of the early *in vitro* embryo, disagreements it did not want to bring to the broader public's attention. New therapies that might actually help daughters and sons, fathers and mothers, wives and husbands, might cause millions of people to examine more carefully their beliefs about those earliest *in vitro* embryos destined, in any event, never to be implanted in a womb. However firm their opposition to abortion, they might discover distinctions that would permit them to support stem cell therapies.

Last and most immediately, scientists can move aggressively to carve out on the slippery slope to reproductive cloning a firm and level place on which to stand. What can the scientific community do to deny to immoral and irresponsible would-be cloners the means to pursue their indefensible goals (Murray 2001)? Respected scientists have repudiated the unfounded empirical claims and denounced the declared intentions of people like Boisellier and Zavos. What more can be done to expose them? To prevent them from obtaining the materials and expertise they need to pursue their goals? Are there ways to focus research using nuclear transfer in stem cells to maximize its usefulness for therapeutic cloning while not greasing the slope to reproductive cloning? This is the challenge to scientists. I can only pose it and show why it is important to rise to it. It will be up to the scientific community to take it on or not. Whatever the community's response, I hope that I have made the implications for ethics — and for politics and policy — clear.

References

Public Health Service Act 2003 A bill to amend the Public Health Service Act to prohibit human cloning. United States Senate, Bill No. S.245

Committee on the Judiciary 2003 Promoting ethical regenerative medicine research and prohibiting immoral human reproductive cloning. Committee on the Judiciary, United States Senate. Washington, DC *http://judiciary.senate.gov/hearing.cfm?id=622*

Bush GW 2001 Remarks by the President on stem cell research. Washington, DC *http://www.whitehouse.gov/news/releases/2001/08/20010809-2.html*

Hoffman DI, Zellman GL, Fair CC et al 2003 Cyropreserved embryos in the United States and their availability for research. Fertil Steril 79:1063–1069

Kass LR 2001 Preventing a brave new world: why we should ban human cloning now. New Republic, p 30–39

Murray TH 1996 The worth of a child. University of California Press, Berkeley, CA

Murray TH 2001 Even if it worked, cloning wouldn't bring her back. Washington Post. Washington DC, section B1

National Bioethics Advisory Commission 1999 Ethical issues in human stem cell research. Vol. 1. Report and recommendations of the National Bioethics Advisory Commission. Rockville, Maryland, USA

National Institutes of Health Human Embryo Research Panel 1994 Report of the Human Embryo Research Panel. National Institutes of Health, Bethesda MD, USA

National institutes of Health Office of the Director 1999 Memorandum from Harriet Raab, DHHS General Counsel, to Harold Varmus, Director NIH. Department of Health and Human Services, Washington DC, USA

Office of the White House Press Secretary 1994 Statement by the President. Office of the President, Washington DC, USA

Parens E, Knowles LP 2003 Reprogenetics and public policy. Reflections and recommendations. Hastings Cent Rep 33:S1–S24

DISCUSSION

Gearhart: You mentioned the Brownback bill. For those of you not in the USA, I thought I'd comment on one aspect of the bill. This bill, in addition to banning all cloning or SCNT, proposes that any US citizen leaving the USA to get a therapy derived from human ES cells in another country would be imprisoned and fined on their return. You made the point of saying that we should keep our heads down and work at getting useful results that would convince people to change their mind.

Murray: I didn't say that people should keep their heads down; rather that they should stay focused on their work.

Gearhart: In conversations that I have had with policy makers about what it would take for our current federal administration to change the policy of ES cell research, the answer invariably is to cure someone. We know how many years this is going to take.

Weissman: I was head of another panel that looked at the issue of reproductive cloning. We felt we had solved the whole slippery slope argument by saying that we would call for a legally enforceable ban on reproductive cloning, defining 'reproductive cloning' as implanting into a uterus a blastocyst developed by nuclear transfer technology. The reason I believe this stops the argument cold is that all of you who work in *in vitro* fertilization clinics know that implanting a blastocyst can't be a solitary event by a rogue postdoc. It is a clinical procedure in a clinical environment that requires anaesthesiology, usually gynaecology and the proximity of an urgent response unit of some sort. The conspiracy to commit a crime could not possibly happen in a laboratory where a rogue graduate student would flip and transfer blastocyts into someone's uterus. The whole set of things is so ridiculous that we felt it was adequate to ask for a legally enforceable ban. We didn't ask for it for ever: this is a different set of issues. We took as our standard that if it ever is as safe and effective as IVF we ought to look at it again. It is clear and I don't think there is the possibility of a slippery slope if you call for a ban on reproductive cloning. Where there is a problem is that if people ever did this other than for self-attention and moneymaking, the absence of a ban on reproductive cloning in the USA would permit them to do it. By all of the animal experiments we looked at, the most likely outcome would be a disaster — certainly for the fetus and most likely for the mother too. There were plenty of reasons to ask for this legally enforceable ban because people simply don't listen.

Trounson: Probably the easiest outcome to achieve at the moment from reproductive cloning might well be a child rather than a cure involving ES cells. This is the danger. It is clear that we ought to strongly ensure that the first outcome isn't a cloned child but rather a person cured by ES cell therapies. The report of a cloned child would set back the field for a long time, and we need to be vigilant to stop anyone whatsoever from making a cloned child.

Murray: I think that is true. Leon Kass describes the industrial scale production of human embryos with doctors secretly implanting cloned embryos, and he suggests that this is simple enough that its final steps could be self-administered by the woman. I checked this out with some people who do IVF and they said that this is probably not correct. Actually, they said something stronger than this. This is the sort of 'factual' assertion that is out there. There is an obligation on us not to let these mis-statements go unchallenged.

Nadal-Ginard: I think we are discussing several issues at the same time here. I am not sure about the significance of separating therapeutic cloning from reproductive cloning. The people who oppose cloning tend to be equally opposed to both sorts of cloning, on the basis of the fact that a human embryo is created. In their view, people are being destroyed. The outcome is irrelevant to these people.

Murray: I want to be as fair and accurate as possible here. Clearly the reason that the religious right has been so vehemently opposed to the selective ban that Irv Weissman has proposed is that in their view this means that people will be able to make a baby only if they promise to kill it. This is how it seems to them.

Nadal-Ginard: If it was a rational discussion they would have bought the argument that Irv has presented, but this isn't what the fight is about. The fight is about creating human beings to be destroyed. To take another tack, I strongly believe that in the USA the right, left and centre are never going to ban IVF. IVF is here to stay. This presents the strongest argument to bypass this debate. There are already 400 000 'people' (that is, human embryos) in the freezer. What is going to happen to them? Why isn't there an outcry about this? It's a lot of people. Why is a person made by IVF less of a person than one made by nuclear transplantation? The IVF-generated embryo is more likely to become a person than an embryo made by somatic cell nuclear transfer. Why don't you use this argument?

Murray: Your reasoning is quite good here. That is why in the USA there is not a ban against creating embryos *per se*. This is also why we have no regulation of these clinics, because this would involve official government permission to create embryos. We don't give permission; it just happens because the government has chosen to look the other way.

Trounson: You can't use those embryos in the USA if you have NIH funding.

Murray: That is true. This is what I was explaining about the Catch 22 in policy history. You could do the research with federal dollars if you could get the approval of a committee that didn't exist. It was a clever public policy but it turned out to be politically unsustainable in the wake of Roe v. Wade to have the Federal government officially giving its imprimatur to research on human embryos. Bernando Nadal-Ginard is right: some right to life campaigners do object to IVF and so are consistent in this regard.

Nadal-Ginard: But they have had no success.

Murray: That's right, and they don't publicize this issue because it is a lost cause and it would make people rethink the issue more generally. What's wrong with creating embryos for people who are trying to have a child? Isn't that what life is about? These are some of the things they might be led to think.

Wilkins: I disagree with Bernardo. I don't think scientists should embrace the idea that there are 400 000 people sitting in embryos waiting to emerge. There needs to be a distinction between an embryo that has a potential to become a person and an actual person. This is a fundamental issue that science is going to have to engage with.

Gearhart: Recently, I debated with a member of the President's Council on Bioethics who commented that he was the same person now as he was when he was a fertilized egg. This individual believes firmly that the person doesn't come out of an embryo, but that the embryo is actually the person.

Nadal-Ginard: To use this argument the scientist doesn't have to say that these are people. We can say that 'you say they are people, and you allow them to be in the freezer'.

Murray: To be consistent, one would have to take that position. In fact, there is the 'Snowflakes' movement in the USA where people do so-called embryo adoption. They volunteer to have frozen embryos implanted.

Jaenisch: One of the problems here is equating cloned embryos with fertilized embryos. There is a basic difference. One is that cloned embryos are not new life: there is no recombination of genes. Rather, it is propagation of existing life. Biologically this is a significant difference, even if it doesn't carry much weight in ethics debates. The second difference is perhaps more important and concerns the potential of these embryos to make a normal baby. This is pretty high for an *in vitro* fertilized embryo, but non-existent in practical terms for a cloned one. There is no way to predict which cloned embryos will be normal and which ones won't. If we propose making an ES line from a left-over IVF embryo, this has ethical problems because it involves destroying a potential normal human life. I don't think there is the same problem working with a cloned embryo.

Evans: There is the concept of ensoulment at fertilization. This has been used in arguments.

Wilmut: At some point in the future cloning will work, and then the argument over safety will no longer hold. We have to draw careful distinctions between the safety argument and the longer-term question: even if it were safe, would it be an acceptable thing to do? I would argue that it would not be acceptable because most people would find it difficult to treat a cloned child in exactly the same way that they would treat a child produced by normal reproduction. Although we understand in abstract terms that the clone is a new individual most of us would expect the child to be very similar to the original. This would impose unreasonable expectations and limitations upon the cloned child.

Jaenisch: I think there is a technical problem and a principle problem in the science. The technical problem is to make clones more efficiently. The principle problem is to make abnormal clones more efficiently. The principle problem is simple: if we short cut gametogenesis and begin to equalize the epigenetic difference between the sperm-derived and oocyte-derived genome, you cannot recreate that unless you physically separate the two genomes in the somatic cell. We are so far away from this we don't need to consider it.

Trounson: This reminds me of the pre-embryo debate that occurred with IVF. Dolly was an embryo and if a child was born, they would have been derived from an embryo. The pre-embryo argument didn't work for human IVF and I expect it will not be workable to describe nuclear transfer embryos as something else.

Weissman: About 8–10 years ago I was talking in a scientist–scientist meeting with a bunch of chemists and astrophysicists and so on. I talked about stem cells and they asked me about genetic modification of the germline, and whether we were going to create new people. I said this was ridiculous. One of the scientists said that he was an astrophysicist and the smallest unit of time he studies is 10 000 years. Rudi, are you telling me that we won't have solved the problems of nuclear transfer in 10 000 years' time?

Wilmut: There is an issue here in relation to stem cells. Why are we so enthusiastic to put Rudi's abnormal cells into patients?

Jaenisch: It works! This is exactly the confusion: what are stem cells? I think they are the result of selection for proliferation. If we get stem cells from normal embryos there is no difference between that and a stem cell coming from a cloned embryo. Both go through a selection process. They forget their epigenetic history. If you have blastocysts the cells know exactly where they came from. If you go through this bottleneck of selection the cells forget this. There is no evidence at this point that there is a difference between an ES cell derived from a cloned embryo or from a fertilized embryo. Scientifically, to argue that because clones are abnormal that cloned ES cells are also abnormal doesn't make sense.

Evans: You can make perfectly normal animals from ES cells.

Jaenisch: Not by cloning; only in chimeras. Chimeras have the helper cells so it is a different story. A chimera is not a clone.

Nadal-Ginard: Now that stem cells can generate oocytes, let's assume that they would be able to make a mature oocyte and also mature sperm. If you put the two together you might get an embryo that is not produced by cloning but instead by fertilization, made from two ES cells. Where does this embryo stand in this slippery slope that we were discussing?

Murray: I would be interested to see how the people who view embryos as persons would respond to this. If they are to be consistent, someone like Orrin Hatch, who is a Mormon, would say that *in vitro* embryos are not full persons so it would not be a problem for him. For people who are staunchly committed to the embryos are people thesis, I suspect they would conclude that this too is a person.

Nadal-Ginard: An embryo created in this way has a chance to fulfil the requirements of Rudi Jaenisch: it could be normal.

Trounson: Why would you do it? It wouldn't be compatible with the patient.

Nadal-Ginard: This is a theoretical possibility that probably is likelier than nuclear transplantation, if these sperm and oocytes could be developed. It would be done for reproductive cloning.

Jaenisch: It is not a clone.

Wilmut: The point surely is that this might be a better way of overcoming infertility.

Nadal-Ginard: No, this is for the sake of the argument with people who want to stop this work.

Jaenisch: It is extreme reproductive assistance, but I don't see any ethical problem with it. It is just technology going to a new extreme. If you set up imprinting correctly *in vitro* these would be normal babies.

Murray: I was saying that Hatch would say it is not a person, but it would be if it were implanted.

Weissman: I have talked with him: he says he is pro-life and there is no doubt which side of the question is pro-life here. He does not consider cells that are not implanted as humans.

Nadal-Ginard: This fertilized embryo doesn't create a moral problem then.

Weissman: The moral problem of the Catholic Church is that anything that goes beyond a fertilized egg is to them a human being, with all the rights and potentials. We'll never get over that: they aren't going to change their mind. Although you say this is not an argument between religion and science, in that particular case it is.

Murray: Bear in mind what your astrophysicist friend told you. Change does happen, even if it is slow.

Weissman: The Catholic Church changed their mind about Galileo only 400 years afterwards!

Murray: I am not a theologian, so what I say about theology you should take with a grain of salt. But from what I know, the position of the church has evolved. It didn't always view the earliest stages of life as equivalent to personhood. This was a relatively recent decision. You may have your disagreements with the Catholic Church's stance on this position, but it is a tradition with a long history of high-level intellectual discourse. People are open to argument. What worries me more is the very powerful political force in the USA of the Christian right, which is very anti-intellectual and doesn't have this tradition of careful discourse and dialogue with science.

Trounson: To say it is not a religious issue seems inaccurate to me. We are not saying it is one religion, but the argument is maintained by strong Catholic and some Anglican theologians. If you look at the scientists who argue against the opportunities in these areas, they have a strong religious background. If we remove the morality issues concerning the embryos used for making stem cells, most of them think it would be reasonable to explore cell therapies with all types of stem cells.

Murray: My thesis is probably controversial even in the USA. Let me say a bit more about why I think it is a mistake to see that as primarily rooted in religion. Remember what I said about tapestry and the thread of the status of the embryo woven into images of good lives and the different paths of male and female lives. When I wrote about this it was based less on my philosophical training and more on my experiences as a single parent rearing young children, which was probably the most important thing that happened in my life. I am writing for men who want to be parents, as well as for women who want not to be so constrained by their fertility that they have no other options in their lives. If this is plausible, clearly cultural forces shape religious beliefs in very complicated ways. There is no question that on the Christian right, the views about abortion, embryo research and the nature of men and women are tightly tied together. I would argue that these are at least as much deep cultural views as they are religious ones. Newt Gingrich famously once said that the reason women shouldn't serve on the front line is that they get sick every 28 days.

Trounson: If we said that you could re-programme cells to be pluripotential without using the egg and five days of culture, do you think there would be any objection to ES cell research? My view is that there would be none if there was never an embryonic stage.

Murray: No, there would still be objections. They wouldn't come from the same quarter, but it would come from the Kass-type arguments about slippery slopes, egg manipulation and so on.

Grompe: If you balance that against the therapeutic potential, I don't think there would be an argument.

Murray: I don't think that argument would prevail. People would be more persuaded by the therapeutic argument.

Minger: This is interesting because the church in the UK has been very involved in all the legislation regarding human embryo research. The HFEA panel has theologians on it. The church has come out actively in support of this research. Using the term 'religion' is quite open-ended.

Murray: There are different faith traditions that take different stances.

Weissman: For those people who come to the debate with strong negative points of view, most are coming with a strong religious perspective. Even Leon Kass, who has written a book on the Bible.

Murray: I know his book on the Bible only from reviews. According to the review from the Sunday *New York Times*, it was very much about men and women and how the old testament essentially gives man hegemony over women.

Grompe: The objections are moral–religious and you can't put this any other way. It is not a scientific argument.

Weissman: If we allow this research to be banned then people's lives will be lost. Recombinant DNA research was almost banned in 1981, and how many people's lives are saved every year from the fruits of recombinant DNA research, and what branch of life science doesn't use it now? To those who advocate banning this research, I say that they must take personal responsibility for the lives lost. If they propose a 4 year moratorium, then there will be lives lost over this four year period. At graduation from medical school we said that our highest priority was the health of the patient. Personally, I extrapolate this to be not only the patient in front of me but also the patients we are all working for. We can't let our own personal views stand in the way of this.

Murray: I think the argument about suffering and premature death is the strongest reason for this research. It is incumbent on the scientific community to communicate much more effectively to the public and policy makers the value of nuclear transfer and stem cell research for basic biological research. The focus has been on transplants.

Lovell-Badge: There are couple of problems. The term 'therapeutic cloning' is there, and it is very hard to get rid of it. Another point is that it is not just religious people who object. The Green Party in Germany also object to it. They are pro-abortion but against embryo research.

Murray: There is a small contingent in the USA with a similar point of view, but politically they have little impact.

Solter: I agree that ES cells will be more important in defining basic mechanisms of diseases and finding drugs. However, we do not need to do nuclear transfer and therapeutic cloning for this. We can take the existing ES cells and manipulate them genetically any way that we want.

Weissman: Yes, if you knew all the genes involved in a complex disorder. Why not just get the cells from the patients with the disease? In amyotrophic lateral sclerosis (ALS) there are genes such as *SOD* that are well known and then there are poorly defined groups with candidate genes. In this way you can test each candidate gene.

Lovell-Badge: There are also complex genetic diseases where you'd have to change many genes at once.

Minger: One way you can get around this is to use embryos that have been screened by pre-implantation genetic diagnosis. We have generated a cell line from an embryo that has a cystic fibrosis disorder.

Final discussion

The regulatory environment

Gearhart: In this final discussion I'd like to start by asking participants to give an overview of the regulatory environment in each of their countries.

Solter: In Germany we have two laws. One is concerned with embryos and the other with embryonic stem (ES) cells, the latter being introduced very recently. In Germany the embryo is given the full rights and moral values of a human being from a moment which biologically does not exist: the fusion of two pronuclei, or shall we say the moment that pronuclear membranes are dissolved. It was essential that the fertilized egg in which the two pronuclei are separated is not considered an embryo in a legal sense, otherwise IVF would be impossible. Accommodations have been made with real life. The other accommodation with real life in Germany is that while abortion is illegal, it is not punishable. This was made because there were two Germanies, in one of which abortion was completely illegal and punishable, and in another in which it was allowed. Thus on reunification there was a compromise. With stem cells the situation is that it is not possible to derive stem cells. It is only possible to work with stem cells that were established before 1 August 2000. One can import them with permission from the Federal Ethical Commission, of which I am a member. These cells are not treated the same way as other human cells because there was a feeling that they are totipotent and therefore they might be the same as embryos. Working with them requires more justification than working on any other human cells. This justification has to satisfy the criteria that every single possible bit of research that can be done on non-human ES cells has been done, and that all possible research on adult stem cells has also been done. It is very hard to fulfil these requirements, so in practice we try to determine whether the experiment is reasonable and of potential use and then we allow it. So far, except for one protocol, all requests were approved.

Gepstein: A lot of the early human ES cell work was done in Israel. Of the first 10 papers published on human ES work, six involved Israeli scientists in some manner. The Israeli law on this was passed in 1999, and put a five year moratorium on reproductive cloning and germline modification, but it did not prohibit the generation of stem cell lines. After the big ethical debate that took place after the generation of the human ES cell lines the Israeli academy of science appointed a committee to put out a report advising on the possible use of

ES cells and nuclear transfer. This committee included scientists, bioethicists, religious figures but no politicians. The Jewish and Islamic positions are very close: they believe that there is a distinction between life and the potential for life. An IVF product is not the same as an embryo in the uterus, and there is a moral obligation to do research if this can result in therapies for treating human disease. The report from this advisory committee concluded that they would permit the generation of stem cell lines and research on human embryonic stem cells. They also permit nuclear transfer experiments as long as embryos are derived from supernumerary IVF frozen embryos and that there is parental consent. There is prohibition on the sale of IVF embryos and reproductive cloning. In the last month Israeli law has been revised again, stating that there is a prohibition on reproductive cloning but making a clear distinction between reproductive and therapeutic cloning. Stem cell work is allowed as long as it goes through the local Institutional Review Board (IRB) and the Central Health Ministry Committee.

Soria: Spain is a Catholic country and the law has recently been changed. In 1988 a law on human assisted reproduction allowed research on non-viable embryos. In 1998 the consulting committee for this law recommended that embryo research on remnant embryos should be allowed. At this time it wasn't clear whether we could use non-viable embryos or human ES cell lines derived from viable embryos. The history begins in my case in the year 2000 because after publishing a paper (Soria et al 2000), I applied for a grant from the Juvenile Diabetes Research Foundation. The grant was big enough — one of the biggest ever given in Spain — but I couldn't use the money. I passed the local hospital ethical committee, and the Juvenile Diabetes Research Foundation's ethical committee. But when I said I wanted to use human ES cell lines, the government said no and told me I had to give the money back to the JDRF, and so I began to look for somewhere else to do the experiments. At that time Singapore was the option because both Sydney Brenner and Sir Roy Calne were supporting me. Then things began to change. All the leading Spanish scientists wrote a letter to the government, and all the universities, scientific societies and the General Medical Council joined in saying that this research should be allowed. The socialist party proposed modification of the law eight times. Nothing happened. Scientists and doctors were never able to influence the government's position. However, the Federation of the Association of Diabetic Patients asked for permission to use my name in writing a letter of support. I agreed. I didn't see the letter at the time. A few weeks later, I was in a restaurant with my wife and I was stopped by a girl who asked to sign a letter of support to myself! I said I couldn't sign the letter and the girl was initially upset with me until I explained who I was, and that it wouldn't be right for me to sign a letter of support for myself. I later asked them to stop this letter but by this stage 1.5 million people had signed it in just a few weeks. This is what really changed the situation. The letter was presented to the government, and they named an Ethics

Committee that was completely biased to the position of the Catholic Church, with a majority of Catholics on it, but this committee voted 11–1 in favour of embryo research. This was at the end of 2002. In 2003 the regional government of Andalusia approved a regional law. Spain is an almost federal country. The regions have the capacity to make laws. A few months later the central government modified the law. The situation now is that we can use embryos for research and we can also use ES cell lines. In a week's time we will open a new lab for cell therapy of diabetes in Seville with a budget of 1 million euros. My conclusion is that we scientists don't have the power to change the law. What changes the mind of the government is 1.5 million votes.

Gearhart: So you are now allowed to use human ES cells.

Soria: Yes, and embryos too. The Andalusian law is a little more generous than the central law. We can use any frozen embryo or remnant embryos, but in the central law these embryos have to be deposited in a central facility and cell lines derived under certain conditions.

Trounson: In Australia, until the end of 2003 there were six states and two territories, all with different laws. Scientists travelled around and did their experiments in different parts of the country with different collaborators. The Australian government then opened the issue up to a debate where they removed party lines. This was a difficult time because it allowed people with strong views to make critical statements in parliament where it isn't possible to get redress. In the end, two laws were passed. One banned reproductive cloning and also therapeutic cloning on the basis that embryos were going to be created for research. But this was scheduled to be reviewed three years after the enactment of this bill, which is in 2006. If scientists could make a reasoned case that therapeutic cloning was essential for research or therapies, the politicians would review these acts. Hence, we face another round of discussions with the government. On the other hand, in the other act they allowed us to have access to frozen embryos prior to 5 April 2002. There are around 70 000 embryos in storage. The reason for choosing this date is that they were concerned that clinicians might super-stimulate patients to obtain extra embryos for research. We have access to these frozen embryos providing that we obtain a licence. One of the uses might be for making new ES cell lines. It is imperative that more of them are made. There is a recent report about the US lines: there are only about 12 of these that have been fully characterized and are useful. They registered around 70 lines, but some of them were no more than cells from the inner cell mass frozen in vials. If there are only 12 cell lines available for NIH funded research, that is too few for the best and largest research community in the world. We would like to make at least 20 new ES cell lines and this work will be underway soon. Hopefully we will be able to persuade the government to shift the 5 April 2002 time, because we can't access any genetically abnormal embryos that will be available through preimplantation genetic diagnosis. I think we must begin the argument about the merit of making ES cell lines from patients with genetic diseases and make this a dominant message and the reason for using 'therapeutic' cloning. If anything comes of the present meeting, I hope that it is that we agree to take this forward into our various political arenas. If we do this alone, this meeting will have been worthwhile. I encourage us to focus on this opportunity. The opportunity is not well appreciated for advancement of medical science using ES cells from mutant embryos, and it needs to be articulated at a level that common people can understand. We need to pick some major diseases for which having these cells in the lab will make a potential contribution to a cure. This simple message will undo a lot of the opposition to therapeutic cloning.

Lovell-Badge: Given the rule that you can only use embryos frozen before 5 April 2002, do you have problems with issues of consent?

Trounson: Yes, but the IVF community has been pretty well organized in obtaining additional consents. It would be a minority of embryos in the bank for which you couldn't obtain the necessary consent.

Björklund: The situation in Sweden is similar to the one in the UK. The difference is that nuclear transfer is not yet allowed, but the government has stated that they will make the necessary changes in the law. Interestingly, Norway has a very different regulation. Early this year they passed a bill that is similar to the one in Ireland: they have now banned any research on embryos, derivation of ES cells and also the use of ES cells.

Murray: I was at a meeting where the Prime Minister of Norway spoke. Despite my claims about the connection between religion and opposition to this work being looser than is supposed, he is a minister in a conservative Protestant denomination.

Björklund: I think it is seen as a sign of general conservatism rather than religious belief.

Blumer: Switzerland is an interesting country as it is a basic democracy which implies at any time the potential for a public referendum once a law is made. The Swiss discussions started in 2001, very much following the German discussion.

In 2001, Marisa Jaconi from the University of Geneva applied for an import licence for a human ES cell line. On this occasion the Swiss government discovered that there was a regulatory gap in the existing legislation, and as such an import was not covered. The government offered Dr Jaconi an intermediate permission so she could import the ES cell lines. Later on it adopted a law called the 'Bill Regulating Research on Human Stem Cells'. Unlike the German law it is not named 'Embryo Protection Act', so it is clear that the Swiss government tries to allow research but in a strict framework. In Switzerland the freezing of embryos for IVF is banned, like in Germany. It is only possible to freeze impregnated eggs, before the fusion of the pronuclei. The law was enacted in December 2003 and now votes are being collected for a public referendum with an 8 April 2004 deadline. By then, 50 000 votes will be needed to initiate a public referendum. The experts think these votes will be collected. It will be an interesting year and we need to educate the public. The law defines an embryo as everything with fused male and female nuclei. The creation of embryos for research purposes is banned, as are therapeutic and reproductive cloning. Import of stem cell lines and the creation of stem cells from spare embryos are allowed. But, how do spare embryos exist if freezing of embryos is not allowed? There are rare cases where impregnated eggs are thawed, developed to blastocyt stage and then for some reason—for example, because the woman fell sick or the father suffered a severe accident and family planning was ceased—it is not possible to implant the blastocyst. An estimated 200–2000 of 'spare embryos' exist. Prior to the research use, written informed consent and approval of both parents is required (in cases where the husband died and therefore the embryo was not used, this has not been possible), as well as a proof that the parents received no financial compensation. Additionally, the decision not to use an embryo must have been taken before consent to use the embryo for research. This imposes quite a few hurdles, but for a conservative country such as Switzerland it is still quite a liberal law which even allows the patenting of modified stem cell lines. But at this stage we don't know whether the law will stand up against the referendum.

Trounson: That is more liberal than the Italian or French situation. Switzerland is now a liberal country!

Evans: I'll try to sum up the situation in the UK. 'Therapeutic cloning' is an unfortunate misnomer, and so we have tried calling it cell nuclear replacement (CNR). Reproductive cloning is now illegal: a specific ban is in place. Everything which we do with embryos is covered by the 1990 Human Fertilization and Embryology Act. This is an interesting piece of legislation because it says that everything is forbidden except in pursuance with a licence. The licence has to be provided by the Human Fertilization and Embryology Authority (HFAE; *http://www.hfea.gov.uk*), and they can only give the licence following specific rules. In particular, replacing the nucleus of a cell was considered in the 1990 act and was not allowed. In 1998 a new report was commissioned from the Human Genetics Advisory Commission together with the HFEA. They recommended to the government that there should be a change of the rules under which licences could be given to include two further purposes. One is for nuclear transfer for the treatment of mitochondrial disease (a reproductive cloning) and the other being CNR for research purposes and ES cell derivation. This report was not accepted, but the chief medical officer was asked to set up an expert group to re-assess this subject. This group produced a report that supported CNR but not the mitochondrial disease approach. Following this, there was a proposed modification that gave permission for licences for the use of human embryos to be granted including other purposes, such as therapeutic research purposes. This itself wasn't completely plain sailing: there was a House of Lords committee set up to consider whether we could do nuclear transfer. They spent a year examining this and came back with a report that in essence supported everything but included some extra suggestions for regulation. That is, every licence granted for making ES cells should include the condition that those ES cells must be deposited within a central stem cell bank. This stem cell bank will be responsible for distributing them and will only give them to researchers with the appropriate ethical permissions, and who have been cleared by the stem cell bank steering committee. I am one of the few scientists who sit on this committee. It is chaired by an able clinical scientist , but there is a lot of input from professional ethicists and other interested parties. There is a lot of pressure to regulate the actual use of cell lines in research.

I have always tried to argue forcibly that the fully and properly regulated derivation of human ES cells is clearly required; so also will be proper regulation for their application to any human subject but I do not consider that their research use, particularly *in vitro*, should be regulated.

Lovell-Badge: You are considering cell lines that will be made in the UK and deposited in the bank. What about cell lines that have been made elsewhere?

Evans: They almost certainly fall under the same regulations, although it is a bit of a grey area. I think what we'll find is that any funding to cover the use of these lines will have a clause in it that they must go into the stem cell bank.

Minger: I am not sure about that. You can import any line, but you do have to register that you have lines from abroad. The difference is that we can freely distribute lines that we haven't made, but we can't give out our lines without going through the bank.

Evans: I think some of the regulators on the bank have spotted that loophole and are working on it.

Weissman: We've discussed the USA already. There is a stalemate: there is no law against reproductive cloning or nuclear transfer research at the federal level. There is an executive order that prohibits any US government funding for any cell lines not on the list as of August 2001. In September 2002 California passed a law which made it illegal to produce reproductive clones. It encouraged nuclear transfer research or new ES cell line research but placed it under a regulatory commission that has never been appointed. The bill was passed, but funding was not forthcoming. In response to that, a citizens' advocate group developed an initiative and constitutional amendment for funding all types of stem cell research, with the priority being those kinds that are not funded by the federal government. They have asked for US$3 billion over 13 years. The constitutional amendment was for it to be in campuses other than the University of California. It would have a state stem cell and regenerative medicine oversight group, and the only reviewers of either facilities or grant applications could be people, not in the state of California, but who were experts in the field, and they could have no connection with the companies that might apply. This is a very different attitude. Those who have talked to the new government believe they are in favour of this kind of stem cell research. In several different states there are different things happening. Several states have banned it, New Jersey has passed it, it is pending in Massachusetts. Each of these different notions of stem cell research will be tested out at the level of the states. I met with several people including the head of the NIH, Elias Zerhouni, about two months ago. We went through a number of scenarios. It seems he got the point about the use of cells lines from people with genetic disorders. There are a couple of important points here. Do we need a facility that is clean of any federal funding? The answer is no. Even my own lab could do nuclear transfer research on humans along with NIH-funded work so long as I could show that the funding for the nuclear transfer research covered all indirect costs as well as the direct costs. This is positive. On the negative side, if a cell line is made in the UK or Boston, for example, even if it is proven to replicate a disease, no US funding can be used by any researcher for further investigations. The head of the NIH understands the paradox that in his own institution, there are people who have devoted their lives to research a particular subject who will not be able to use the cell lines that might offer the best opportunity for understanding that subject. I have another point about validating genes which are involved in disease, something that seems to have been missed by big pharma. A lot of work is done validating particular genes and proteins as being targets for small molecules for chronic therapies. I can't think of a more direct way to validate targets in multigenic diseases than to have the cell lines there. It is inappropriate that big pharma has largely stayed out of this debate. If there is a second thing we could do here today, it is to alert big pharma as to which side they should be on. As soon as you start banning research you are putting yourself back in the time of Lysenko in Russia. We have an example of what happened when genetic research was banned in Russia in the 1930s: Russia stopped producing good science in this area for at least 50 years. They missed the recombinant DNA revolution and biotech. It is a good example to put before our own legislative bodies. This research is moving forward. However you think the issues will come out, someone will get a cell line, which will be a big step. If these cell lines exist and you can't work on them, there will be other places to work on them.

Lovell-Badge: I agree with you. This research will allow us to look at human cells in culture rather than experimenting directly on humans. In Europe we have a situation where there are many countries with different views. There are funding agencies that cover all of Europe, notably the EU. There was a moratorium on funding research in this area. They failed to reach an agreement, the moratorium ran out and they couldn't reinstate it, so now the EU is able to fund research in this area.

Solter: I don't want to be negative, but I am afraid that we are going to make the same mistake again and move the goalposts further. We promised that if we were allowed to do therapeutic cloning we would cure people; now we say that we won't cure people but we will research the causes of diseases and create models. Do we have any examples of this? Let's not promise too much.

Weissman: Absolutely. There is too much hype about therapeutic cloning when we don't even know how to mature the cells from an ES cell line in the mouse to get perfect repopulation of the haematopoietic system. Of course, we have to be the voices of reason and restraint about what the possibilities are, but I would argue that if we don't allow people to make a start, progress will be hindered.

Gearhart: This has been a wonderful session. In bringing this meeting to a close, I'll just say this. It is going to take us years to get ES cells to behave in culture in a way that is safe and reproducible. This is where much of our focus is. The other message is the issue of learning what these diseases and disease targets are, and whether cell based intervention will really help. The more I learn about some of these diseases the more of a challenge I realise it will be to get cells in there that are going to function to replace those that have been lost. I am concerned about how effectively we will be able to do this from the outside, and I subscribe to the idea that eventually we will be able to take the patients' own cells and get them to behave the way we want them to. This gets us into another area which I think the ethicists have avoided, which is that if we succeed in this we will have begun to instruct our own cells. This is more powerful than anything we have talked about up to this point. It has been a wonderful discourse and I appreciate you taking the time to attend.

Reference

Soria B, Roche E, Berna G, Leon-Quinto T, Reig JA, Martin F 2000 Insulin-secreting cells derived from embryonic stem cells normalize glycemia in streptozotocin-induced diabetic mice. Diabetes 49:157–162

Index of Contributors

Non-participating co-authors are indicated by asterisks. Entries in bold indicate papers; other entries refer to discussion contributions.

Subject Index

D

E

X

Z